Participatory Biodiversity Conservation

Cristina Baldauf
Editor

Participatory Biodiversity Conservation

Concepts, Experiences, and Perspectives

Springer

Editor
Cristina Baldauf
Departamento de Biociências
Universidade Federal Rural do Semi-Árido
Mossoró, Rio Grande do Norte, Brazil

ISBN 978-3-030-41685-0 ISBN 978-3-030-41686-7 (eBook)
https://doi.org/10.1007/978-3-030-41686-7

This Springer imprint is published by the registered company Springer Nature Switzerland AG
The registered company address is: Gewerbestrasse 11, 6330 Cham, Switzerland

Foreword

Biodiversity conservation is one of the current major environmental problems globally, perhaps one of the top two, along with climate change. As with climate change, biodiversity conservation has a global assessment panel called IPBES (since 2012) to look at all aspects of the problem. The history of conservation is curious. Until recent years, loss of biodiversity (variation of life at the levels of genes, species, and ecosystems) was considered a biological/ecological problem requiring technical solutions. Good science is obviously important, but biodiversity loss is not a technical problem. It is largely a social/cultural/political/economic problem.

For years, I have been fascinated by paradigm changes or conceptual shifts in the study of the environment. These paradigm changes include replacing reductionism with a systems view of the environment to facilitate the examination of relationships and their dynamics. Ecosystems are not predictable, and there is no such thing as "balance of nature"; ecosystems are changing all the time, necessitating that we recognize ongoing change and uncertainty. We cannot "manage" ecosystems or "preserve" them in a static way. We can carry out conservation and try to make environmental systems resilient, with full recognition of constant changes and the variety of drivers behind them, from climate change to global markets.

A second paradigm change in the study of the environment concerns the necessity of including humans in the ecosystem. For many years, the science of ecology (with a few exceptions) assumed away humans. This is an outcome of separating nature and culture, as Gregory Bateson and others have explored in some detail, and is related to the development of a positivist Western science since the seventeenth century. Are those kinds of assumptions and positivist science obsolete? Almost all ecosystems in the world have humans in them, and their activities, practices, and culture impact ecosystems in a variety of ways. To take a holistic view, all ecosystems are really social-ecological systems.

A third paradigm change is the rise of participatory approaches in various environmental fields. We find increasing levels of citizen participation in planning and environmental assessment. We find co-management (sharing power and responsibility for decision-making) in forestry, fisheries, wildlife, and protected area management. In the early days of the "Yellowstone model" of protected area planning, communities were simply removed from a park area, by the use of force as necessary. This happened not only in the United States but also in Africa, Asia, Latin America, Canada, and Australia (parts of Europe were an exception). However, a great deal has changed since the 1872 Yellowstone experience, and conservation has been redefined broadly as more inclusive and participatory.

What do these paradigm changes mean for protected area planning and practice?

First, we see the rise of indigenous knowledge, or traditional ecological knowledge, or, more broadly, indigenous and local knowledge (ILK, as used by IPBES). The use of ILK balances and humanizes positivist science. It facilitates community-based conservation, biocultural approaches, and attention to livelihood needs of local people. Contemporary conservation is no longer solely focused on wild species and wild places; communities are now often central to conservation projects.

Second, and related to the first, including humans in the ecosystem not only provides a fresh perspective, but it also makes it possible to ask new questions and bring in new concepts, enriching the search for solutions. William Adams once commentated that most conservationists are trained to know about biology, not capitalism. As such, they tend to focus on the immediate drivers of biodiversity loss as a matter of urgency, not on root causes. The new questions include those regarding agency, power, and empowerment. We see a global trend of democratization of environmental decision-making, including protected area management.

Third, flowing from the second, we now find democratization examples everywhere: recent protected areas in Africa, such as Namibia's conservancies, are managed by communities for both conservation and community benefits. Australian indigenous people are now able to decide for themselves if their lands would be managed as a protected area and whether the government should be invited as a co-management partner. Canadian national parks in areas traditionally used by indigenous peoples have management boards with community voice and power. Brazil has indigenous lands and legislation that mandates local participation and co-management in protected areas.

This book contributes to a deeper understanding of the issues around social dimensions of conservation, not just for biodiversity but also biocultural diversity. It gives voice to perspectives beyond the Yellowstone model through discussions of cultural landscapes, food security and sovereignty, biocultural restoration, and co-management. Participatory conservation is a theme that runs through the volume, illustrating the relationship between environment and democracy. Cases in chapters are international, but the emphasis is on Brazil, serving as an exemplar for good practice (e.g., *Arapaima* conservation) but also as a cautionary tale for all the things that can go terribly wrong.

University of Manitoba
Winnipeg, MB, Canada

Fikret Berkes

Preface and Acknowledgments

One motivation for producing this book came from the musician and songwriter Paul Simon – but not because of one of his well-known musical compositions, rather from a note published about his tour in support of the "Half-Earth" initiative headed by the eminent biologist Edward O. Wilson. In an interview for the Mongabay site in 2017, Paul Simon stated that he would donate all of the profits from his tour that year to a project designed to designate half of the planet's surface to biodiversity conservation. He said, on that occasion: "We've been talking about what he might want to use that money for… Ed said the Southeast of the United States is the most biodiverse area of the whole country... So he was saying it would be tremendous if we could protect a large swath of area through Alabama down through the Gulf Coast, including Florida. That would be a great achievement. We can explain that to people — say this part is vulnerable and we need to save it."

The good intention of that expressive donation, as well as the importance of spreading an environmental theme among his millions of fans, is unquestionable. Music can be considered one of the arts that has most influenced human societies – a universal language capable of arousing emotions and influencing human behavior. However, for the millions of people who do not go to musical concerts but depend on natural resources for their livelihoods, to simply say that a natural area is vulnerable and must be saved could be quite disruptive and represent just another example of "green grabbing" – land appropriations for environmental purposes that are causing innumerable social impacts around the globe. Nonetheless, even today, so-called traditional conservationists find it difficult to accept (and incorporate) criticism related to negative social consequences resulting from the establishment of protected areas or other top-down conservation initiatives, like Half-Earth, that draw on support not just from artists but also from many scientific communities, organizations, and donors.

Within that context, the present book hopes to give voice to other perspectives related to biodiversity conservation beyond the "fortress conservation" model and emphasize one of the pillars of democracy – popular participation. Publications emphasizing such an approach are not at all novel, as critics of the hegemonic conservation model have existed for various decades, and excellent books on that subject have already been published – as well as innumerable scientific articles and technical documents. Nonetheless, even though the importance of "human dimensions" are now well-recognized within the conservation community, significant numbers of human rights violations still occur throughout the world due to the establishment of protected areas, as was documented recently by the United Nations Special Rapporteur on the Rights of Indigenous Peoples.

In response to historical tragedies that have befallen indigenous peoples and local communities (IPLC) and to the "sound of silence"[1] of many defenders of fortress conservation in relation to those problems, we have witnessed the growing recognition in recent decades that other models of biodiversity or commons conservation must be considered. Those models are based on concepts and approaches such as adaptive co-management, biocultural conservation, community-based conservation, participatory biodiversity conservation, participatory action research, and right-based approaches, all have as their central theme the participation of IPLC in the administration of their territories and the biodiversity contained therein. However, as noted by Elinor Ostrom, participation, in itself, is not a panacea, as there is no single and simple solution for the complex problems related to the sustainability of socio-ecological systems. As such, we present here experiences covering distinct aspects of participative biodiversity conservation and related themes, discussing their benefits as well as their limitations and difficulties.

This book begins with reflections concerning the contemporary crisis of representative democracy, which has revealed the necessity of participatory mechanisms in various fields, including biodiversity conservation (Chap. 1). As already mentioned, however, that participative perspective is not a consensus within the conservation community, as differing visions still divide it. Chapter 2 will therefore explore alternate models as well as current perspectives for biodiversity conservation, as well as the ethical precepts, conflicts, and points of view held in common. Chapter 3 presents one of the most controversial and intriguing topics in conservation – the existence of cultural landscapes in areas otherwise considered pristine by many conservationists. Those authors explore, based on an extensive bibliography, the expansion of mosaics with different degrees of landscape domestication during the Holocene and reflect on the importance of involving human populations in the conservation of landscapes that they themselves generated.

Among the diverse social impacts caused by the fortress conservation model, emphasis is placed on those related to food security and sovereignty. To that end, Chap. 4 presents a review of the importance of wild foods to human populations, especially among low-income groups, and reflects on the impacts of protected areas on access to those food resources (as well as other forest resources essential to local livelihoods). Chapter 5 complements that investigation by discussing conciliations between food production and conservation and presents the results of

[1]Art Garfunkel summarized the significance of that song as "the inability of people to communicate with each other."

10 years of participative processes linked to livelihoods, agroforest systems, and conservation in a MAB reserve in Mexico.

The following three chapters present case studies involving the management, conservation, and participative restoration of biodiversity. Chapter 6 focuses on the cultivation of microalgae that combines income generation, biodiversity conservation, and empowering women in coastal communities in the semiarid region of Brazil. Chapter 7 presents the results of community-based management experiences, with emphasis on two large-scale programs in the Brazilian Amazon. The first was designed to recuperate stocks of the world's largest freshwater fish (*Arapaima gigas*) and thus guarantee their conservation; the second focused on the recuperation of giant turtle populations. Based on those experiences, the authors discuss the possibility of large-scale biodiversity asset co-management in the Amazon. Chapter 8 deals with an action-research project that involved a research group and an indigenous civil organization and was designed to promote the conservation and productive restoration of landscapes and establish agroecological systems in Mexican indigenous communities.

The diverse participative approaches discussed in those texts are frequently targeted with criticisms related to the lack of efficient evaluations of the methodologies applied. In that sense, the three chapters set forth important reflections designed to overcome those doubts. Chapter 9 is an invitation to reflect upon the limitations and shortcomings of planning and participative practices in protected areas. That call for reflection is based on the systemization of various experiences in Spain and focuses on two distinct and complementary aspects: internal problems and external influences. Chapter 10 discusses the role of local perceptions in environmental diagnoses and presents methodological pathways more suited to understanding human perceptions as related to natural resources, without downplaying their intrinsic complexity. Chapter 11 evaluates the participation of IPCLs in five Indigenous and Local Knowledge (ILK) conservation initiatives in the Amazon basin and presents strategies to optimize the roles of those actors in biocultural conservation. The authors also elaborate on the importance of those people to the conservation of the Amazon forest – exactly at a moment when a conjunction of Brazilian political forces is providing incentives for the degradation of the world's greatest tropical forest and ignoring the persecution of its original guardians.

The role of IPLCs in forest ecosystem conservation is also discussed in Chap. 12, which develops the vision of "environmentalism of the poor." The author of that chapter argues for support for IPLCs, especially the women, who represent a fundamental component in the implementation of policies to avoid deforestation such as the REDD+ initiative (reducing emissions from deforestation and forest degradation). Finally, Chap. 13 attempts to summarize the essential findings of the earlier chapters and offers its readers a synthesis of the challenges and opportunities presented by the different facets and approaches of participative biodiversity conservation.

In conclusion, I would like to thank all of the authors who believed in this project and found time to prepare their chapters and share their experiences. I also thank João Pildervasser (Editor, *Life Sciences*) and Luciana Christante de Mello (Editor, *Life*

Sciences) for their invitation to organize this book, as well as the anonymous reviewers whose suggestions were crucial to improving the initial proposal; Sanjana Sundaram (Production Editor) and, once again, João Pildervasser for their support during the execution of this project; CNPq for the research productivity grant (process 308628/2016-0); as well as my students Francisco Lidiano Oliveira, Ivinna Kariny da Costa Vieira, Kevyn Rodrigues da Silva, and Raquel Bruna de Lima for their technical support. Last, but no less important, I thank my husband Miyoshi (Saka) – a full master of participatory parenting – and our little Martina who was born during the organization of this book and gave new significance to my desire for a better future for both humans and nonhumans.

Mossoró, Rio Grande do Norte, Brazil — Cristina Baldauf

Contents

Author the Editor

Cristina Baldauf has a BSc from Universidade Federal do Rio Grande do Sul, MSc from Universidade de Santa Catarina, and PhD from Universidade Estadual de Campinas. She has worked as an External Consultant for the Center for International Forestry Research (CIFOR) over the years, particularly on an initiative developing a research strategy for global dry forests. Currently, she is Associate Professor at Universidade Federal Rural do Semi-Árido in Brazil, where she leads the Ethnoecology and Biodiversity Laboratory. Her research explores the trade-offs between traditional and local livelihoods and conservation using various approaches and theoretical frameworks, particularly participatory conservation.

Part I
Introduction

Chapter 1
Reflections on Democracy and the Participation of Society in Biodiversity Conservation

Cristina Baldauf

1.1 The Contemporary Crisis of Representative Democracy and Its Relation to the Biodiversity Crisis

> I feel that it is necessary to continue to believe in democracy, but only in true democracy. When I say that the democracy in which we live today in this world is a fallacy, I am not attacking democracy, far from it. It's just that what we call democracy isn't. And, if it ever is, the difference will certainly be perceptible– José Saramago

The word democracy dominates our perceptions today, and it is extremely rare that governments or societies do not self-proclaim themselves as democracies. If, on one hand, democracy is defined as government by the people (following its rigorous etymological origin), it must be recognized, on the other hand, that the term assumes many different connotations. The pillars of democracy are embedded in the supremacy of popular will, equality of rights, and the preservation of liberty. The basic functioning principle of contemporary democracy is representativity: the right of citizens to participate in questions of collective interest through their vote, with the principal function of voting being to choose representatives (Medeiros and Albarado 2013).

There is a well-established consensus today on the existence of a crisis in the model of representative democracy. A survey undertaken by the Pew Research Center interviewed more than 30,000 people in 2018 and revealed that 51% of the interviewees were not satisfied with the form of democracy currently functioning in their country. Among the themes that stood out, within that survey in terms of the general discontentment with democracy, were economic frustrations, the status of individual rights, and the perception that the political elites are corrupt and not at all concerned with the fates of common citizens (Wike et al. 2019). It is important to

C. Baldauf (✉)
Departamento de Biociências, Universidade Federal Rural do Semi-Árido, Mossoró, Rio Grande do Norte, Brazil
e-mail: crisbaldauf@ufersa.edu.br

C. Baldauf (ed.), *Participatory Biodiversity Conservation*,
https://doi.org/10.1007/978-3-030-41686-7_1

remember that the political scientist Steven Levitsky, one of the authors of the book *How Democracies Die*, emphasized that democracy is threatened whenever the political establishment loses the confidence of its citizens.

The traditional political system represented by parties has, in fact, failed, as most representatives do not respond to the demands of society because of the monopolization of power by private economic interests and specific groups. As such, the representative democratic electoral system is not, by itself, capable of guaranteeing the general interests of the people. Additionally, the most vulnerable social groups and ethnic minorities face significant difficulties in guaranteeing that their rights are represented. Consequently, that "vacuum" in democracy is expressed through increasing inequality, unemployment and/or precarious work situations, the expansion of zones where the state does not act (and where urban violence and violence against women are the norm), oligopolistic control of means of communication, the absence of agrarian reform, growing environmental degradation, increasing spending on repression, raising fascist governments, and the reliance on war to resolve conflicts—among many other problems—that result in a recurrent question: is it possible that democracy and capitalism are compatible? (Monedero 2012; Fig. 1.1).

Numerous authors have studied the multiple dimensions of the crisis of representative democracy, and it is not my objective to delve into that theme, but rather to attempt to establish connections between that crisis and the decline of biodiversity, or, better yet, the decline of biocultural diversity—for the inseparable nature of biological, cultural, and linguistic diversity, which is now recognized. Innumerable articles and case studies could be used to demonstrate the connections between those two crises, but I will stress here one that is extremely familiar, urgent, and current—the weakening of democracy and its relationship with the loss of sociobiodiversity in Brazil.

Despite the recent pronunciation of the 74th General Assembly of the United Nations that the Brazilian government should work to defend the natural environment, many decisions by the current President Bolsonaro have been to the contrary (Fig. 1.3a). Soon after assuming the reins of government, the President considered extinguishing the Environmental Ministry (only reversing that decision after being pressured by businessmen and politicians linked to agro-business who were concerned about the global image of Brazil and possible European commercial barriers). As it was not possible to eliminate that Ministry, the President opted for weakening it. He began by choosing Ricardo Salles to head the Environmental Ministry, a person whose curriculum included conviction for fraud in preparing a management plan for an environmental protection area when he was the Environmental Secretary for São Paulo State. As Minister, he reduced by almost 95% expenditures destined for climate policy and exonerated the Executive Coordinator of the Brazilian Forum of Climate Change. Additionally, the President declared that climate change was an academic theme and would not have priority in his government; he also declined to host COP 25—the 25th Conference of the Parties of the United Nations Framework Convention on Climate Change (UNFCCC) (Trigueiro 2019). The threats of Ricardo Salles directed toward the Brazilian Park Service (ICMBio—Instituto Chico Mendes de Conservação da Biodiversidade)

Fig. 1.1 Expressions of democracy crises. Inequality triggers mass protests in the United States ("Occupy Wall Street"—2011; **a**) and Chile ("Chile Woke Up"—2019; **b**). A big screen behind the stage in Roger Waters' last tour pleaded to "Resist neo-fascism" and then displayed a list of countries and their respective leaders. (Photos: David Shankbone (**a**), Carlos Figueroa (**b**), and Matheus Paiva (**c**))

likewise led to the president and three directors of that organ to resign (who were subsequently substituted by military personnel with considerably fewer technical qualifications). Additionally, numerous criticisms were voiced by the Minister and the President himself concerning environmental monitoring by the Brazilian Institute of the Environment and Renewable Natural Resources (IBAMA), and its budget was reduced so that the number of fines applied for environmental crimes in the first semester of 2019 was the lowest in 11 years (Terra de Direitos 2019). It is not at all difficult to understand that the success of monitoring and control operations depend on the initial secrecy of those actions—but the communication branch of the current administration recently gave previous warning about important raids. Therefore, in addition to alerting those responsible for illegal lumbering, the government itself exposed its own agents to the risk of being the targets of programmed attacks by armed gangs involved in illegal wood harvesting in the Amazon region (Trigueiro 2019).

In order not to tire my readers' eyes, I have initiated a new paragraph here, but, unfortunately, attacks on Brazilian democracy and socio-biodiversity have not terminated or even decreased. It is absolutely necessary to point out that the numbers of military officers in the upper echelons of government are currently greater than those seen during the dictatorial government of General Castelo Branco (1964–1967) in the initial phase of the military dictatorship in Brazil. The government has also assumed closer relationships with religious fundamentalist groups that now occupy the recently created Ministry of Women, Families, and Human Rights (under the direction of Damares Alves). The agencies responsible for the politics of indigenous peoples, *quilombola* (ex-slave) communities, agrarian reform, and the environment are now in the hands of ruralists and those who oppose the rights of traditional cultures (Terra de Direitos 2019). As such, it is no surprise that within only 10 months, the current administration has liberated the use of 382 different pesticides, with approximately 30% of their ingredients being prohibited by the European Union (Carvalho 2019). There is also a law under discussion in the National Congress that would establish new regulations for the use of pesticides in Brazil, weaken requirements, and transfer their control from the Environmental Ministry and the National Health Agency (ANVISA) to the Agricultural Ministry (Abessa et al. 2019).

Brazil was able to reduce deforestation in the Amazon in the last decade, in consonance with promises made during the Paris Agreement on climate change. Nonetheless, data provided by the *Brazilian National Institute of Space Research* (INPE) revealed record numbers of fires in the Amazon region in July and August, 2019, in relation to previous years, which captured the attention of both the national and international media (Fig. 1.2). In light of that scenario, which in its final analysis reveals the advance of agro-business in the Amazon region, Brazil stopped receiving €35 million in resources from the German and Norwegian governments designed to combat deforestation. The publication of those data from INPE caused considerable anger in President Bolsonaro, who not only questioned one of the most modern monitoring systems in the world, but fired its director, Ricardo Galvão. Another example of denying the validity of the scientific data is the attempt to disqualify research conducted by the immunologist Mônica Lopes Ferreira, who has had an impressive career at the Butantan Institute (a public institution affiliated with the São Paulo State and considered one of the major scientific centers in the world). Her research analyzed the ten most widely used agrochemicals in Brazil and discovered that all of them, at any exposure level, will cause serious impacts on human health. In response to the publication of those results, the Minister of Agriculture gave an interview contesting the data. Ever since then, Dr. Ferreira has suffered persecution within the Institute and has been forbidden to offer courses, participate in seminars, or undertake new research projects (Simões 2019).

Fig. 1.2 Natural-color image of smoke and fires in several states within Brazil including Amazonas, Mato Grosso, and Rondônia. (Photo: NASA)

In addition to opposing important research and firing people with technical expertise and substituting them with military or police personnel without any expertise in environmental areas, one of the most frontal attacks on democracy by the Bolsonaro government was the extinction or fragilization of popular participation in decision-making processes. Provisional Act 870/2019, published on the first day of his government, extinguished important government ministries (such as those of Labor, Culture, and Sports), secretariats, and groups (the Special Secretariat of Family Agriculture, Agrarian Development, the National Commission to Combat Desertification, and the Director Committee of the National Fund for Climate Change), and transferred other responsibilities between different government organs. Decree 9.759, edited in April 2019, resulted in the extinction of more than 700 councils, commissions, forums, and other channels of dialogue between the government and the population. Many of those spaces were related to environmental concerns, but other forums were directed more toward policies and social questions, and were all frontally impacted (Terra de Direitos 2019). Specifically in relation to the National Environmental Council (Conama), a government decree in May 2019 altered its composition (which was previously composed of 96 members that included public entities, social groups, and NGOs) and established a council with only 23 members (with increased numbers of federal government participants; Soares 2019).

Other sets of proposed laws and constitutional amendments that threaten the environment and Indigenous Peoples and Local Communities (IPLC) are currently under discussion in the Congress. The most serious among them are: (1) a law that would eliminate environmental licensing for new infrastructure projects and other economic activities, under the pretext of eliminating barriers to development; (2) a law that would result in the elimination of legal reserves (protected areas in rural properties that cannot be deforested); (3) a constitutional amendment that would prevent the creation of new indigenous lands or conservation areas, and would allow revoking the protection of areas previously determined for those ends; and (4) other laws that would allow the exploitation of water and mineral resources within indigenous lands and protected areas. In terms of the exploitation of natural resources within protected areas, the Environmental Ministry, in opposition to a technical decision by IBAMA, authorized oil exploration activities near the Abrolhos National Park—the site with the greatest known biodiversity in the South Atlantic (Abessa et al. 2019; Trigueiro 2019). In terms of IPLC, the federal government plans to revoke the National Policy for the Sustainable Development of Traditional Peoples and Communities (Abessa et al. 2019) and transfer the capacity to delimit indigenous lands from the National Amerindian Foundation (FUNAI) to the Agricultural Ministry. At least in the latter case, the government's plans have not yet been successful, as those changes were unanimously barred by the Federal Supreme Court (STF) (Abessa et al. 2019; Schreiber 2019).

It is important to note that the repercussions and impacts of all of those actions and intentions extend beyond the borders of Brazil, due to the role of this country in global climate regulation and global hydrological cycles (Strassburg 2019). Nonetheless, meetings between Bolsonaro and representatives of other countries are of great concern to environmentalists. In September 2019, representatives of the Brazilian government met with Mike Pompeo, the Secretary of State of the United States. During that meeting, the Bolsonaro and Trump governments promised to promote the sustainable development of the private sector in the Amazon region and to create a $100 million fund directed toward biodiversity conservation. Although no details of the contract have been put forward, the use of the word "development" in connection with the largest tropical forest in the world has raised considerable concern (Mendonça 2019). In the following month, Bolsonaro met with the Prince (and presumed future leader) of Saudi Arabia, Mohammed bin Salman, and proposed investing resources from that Kingdom to create tourist areas (with resorts inspired by Cancun) that would include an established conservation area in Angra dos Reis Bay in Rio de Janeiro State. To attract that investment, the Brazilian president promised he would revoke the decree that originally created the conservation site (Senra 2019). It appears that Bolsonaro has serious difficulties in accepting the existence of Ecological Stations and their regulations, as he was fined a few years ago for illegally fishing in an Ecological Station when he was a congressman. It was also not at all surprising that just a few months after assuming the presidency, he fired the ICMBio employee who fined him on that occasion. Additionally, he nominated Jorge Seif Júnior as director of the Federal Fishing Agency, even though a company owned by his father had been fined for fishing 12 tons of an endangered species whose trade is prohibited. To resolve that problem, the new director sent a technical note to the

Environmental Ministry requesting the suspension of the list of species threatened with extinction—based on the fact that the list would result in economic losses.

As if the actions cited above were not enough, at the end of August 2019, signals began to appear of what would soon become (according to the Brazilian Justice Department) the largest oil spill yet registered along the Brazilian coast. The investigation into the oil spill that has already affected 2000 km of coastline has been maintained in secret—but in the meantime, many theories about the origin of oil have been considered based on mathematical models of marine currents, conspiracy theories, and pure imagination. But, independent of the origin of that oil, whether it be from a ship sunk during the second world war, a mystery Venezuelan tanker, or a criminal spill caused by Greenpeace, there was one overwhelming consensus: of the inability of the Brazilian government to deal with the crisis. The Environmental Minister, Ricardo Salles, only formally activated the National Contingency Plan (PNC) fully 41 days after the first tar patches appeared on the beaches of northeastern Brazil. The PNC was established in 2013 precisely to deal with emergency situations such as oil spills. In response to that blatant inefficiency, thousands of people living along the northeastern coast worked to remove immense volumes of oil and tar with their own hands—burning and irritating their skin ever since the spill first appeared. Those "volunteers" did not have any alternative except to risk their own health to clean the beaches and fishing grounds from which they earned their livelihoods. That spill has already caused enormous socioeconomic and ecological impacts (Correia 2019).

Almost certainly, between this moment and the publication of this text, other attacks on Brazilian biodiversity and IPLC will have occurred under the false allegation of strengthening the country's economy and sovereignty. Those attacks on nature represent attacks on democracy itself, as both occur against the popular will. A public opinion poll undertaken by the Pew Centre before the Paris Accord revealed that Brazil was the country most concerned with climate change, with 88% of all Brazilians being in favor of national limits on greenhouse gas emissions. Additionally, more recent polls have demonstrated that Brazilians, to a greater extent of any population in the world, desire even higher levels of protection for the natural environment (Strassburg 2019).

Due to the gravity of the current situation, resistance to attacks by the Bolsonaro government has been increasing and involving different national and international sectors. In opposition to the recent liberation of large numbers of agrochemicals, however, there is a Bill circulating in the National Congress that proposes the institution of National Policies for Reducing Agrotoxins, which proposes restrictions to the application of those chemicals and incentives for the transition to sustainable models of agricultural production (Terra de Direitos 2019). The Accounting Tribunal of the federal government has opened an investigation into whether the political environment in the country is affecting the monitoring and prevention of illegal lumbering (Carmo and Neves 2019). At the international level, the reactions that appear to have been most important are related to economic sanctions promised by various European countries if increases in deforestation are not reversed. Pushed by the threats by the French president Emmanuel Macron to block an important

Fig. 1.3 (**a**) In a speech referencing God and patriotism, President Bolsonaro denied the increase in deforestation rates in the Amazon region during the last UN General Assembly in New York. Bolsonaro's unpopular statements and projects have provoked many protests in Brazil (**b**, **c**) and other countries (**d**, **e**). (Photos: Alan Santos (**a**), Romerito Pontes (**b**), Cristina Baldauf (**c**), Cleo Grimaldi (**d**), Grzegorz Wysocki (**e**))

commercial agreement between the Mercosul and the European Union if Brazil were to leave the Paris Agreement, other countries, such as Germany and Finland, have likewise made their opinions clear; the latter country has, in fact, suggested that the European Union study the possibility of boycotting beef produced in Brazil.

Civil societies have also become mobilized in response to the current situation (Fig. 1.3b–e). Thousands of people throughout the world have gone onto the streets to protest the burning of the Amazon. Those movements began on social platforms, with the hashtags #PrayForAmazonia and #PrayForAmazonas reaching approximately 6,000,000 responses, including by celebrities such as Madonna and Cristiano Ronaldo. In its call for prayers, the Catholic Church has also repeatedly manifested its concern about the growing rates of deforestation and of violations of human rights in the Amazon region since Bolsonaro has taken office. A recent meeting of bishops in the Pan-Amazonian region organized by the Vatican had, as its central theme, "Amazonia: new paths for the church and for integral ecology." The working document of that meeting, the *Instrumentum Laboris*, addresses the problems derived from uncontrolled exploitation of the natural riches of the Amazon region, including deforestation, the pollution of its waters by mining, enormous hydroelectric projects, monoculture, and the disrespect of indigenous tribes and their cultures and territories (Kujawski 2019). The indigenous populations in the Amazon, in turn, whose lands and livelihoods are literally burning, have overcome their traditional ethnic rivalries and wars of the past to address a greater cause: the fight against threats to the Amazon region raised by the Bolsonaro government. Indigenous groups that have been historic enemies recently met in the Menkragnoti Indigenous Lands of the Kayapós. One of the results of that event, which united representatives of 14 indigenous ethnic groups, was the creation of a council composed of participating organizations to unify their demands and facilitate political articulations to defend their forests and territories (Fellet 2019).

It is hoped that this long explanation about the contemporary situation in Brazil has demonstrated the absolute link between the erosion of democracy and threats to biodiversity conservation. It is well established that crises, such as the threats to democracy that are occurring worldwide, can result either in the strengthening of institutions or their erosion. Contrary to some observers, who defend the idea that we are on the road to a post-democratic world, the intention of this book is exactly the opposite—as a defense of the democratic ideal—not only as a form of government, but as an exercise of individual participation in collective actions. To that end, in addition to identifying problems originating from barely democratic (or even nondemocratic) approaches to conservation (top–down and "fortress conservation" approaches), the authors of this book suggest approaches, instruments, and effective mechanisms to face the challenges of the democratic conservation of biodiversity.

1.2 Participatory Conservation of Biodiversity: Definitions and Assumptions

In recent decades, so-called top–down models have been widely criticized for not including social dimensions in the conservation of biodiversity, as the creation of protected areas frequently involve the removal of IPLC from their territories and/or the establishment of restrictions on their activities (such as hunting, fishing, extractivism, and agriculture). Additionally, even outside protected areas, the environmental laws of many countries have negatively impacted local ways of life, especially those of rural populations. Changes in political paradigms during the

1960s and 70s, on the other hand, with the end of authoritarian regimes, the consolidation of democracy, and movements in favor of human rights and environmental concerns in developing countries, have resulted in many structural changes in the theories and practices related to the protection of the natural environment: from exclusive to inclusive, and from top–down to bottom–up (Khadka and Nepal 2010).

A number of approaches have been proposed and implemented under the concept of "bottom–up" and "community-based" practices, such as community-based management (CBM), community-based conservation (CBC), community-based environmental protection (CBEP), community-based environmental planning organizations (CBEPO), community-based natural resource management (CBNRM), integrated conservation and development programs (ICD), common pool resources (CPRs) management, adaptive (co-)management, participatory biodiversity conservation, and biocultural conservation. In spite of the fact that those approaches have distinct theoretical origins, they are based on the principle of "participation"—the concept that unifies the different chapters of this book. Participation represents a way of categorizing the levels of control that individuals and groups hold in terms of designing policies in favor of their interests (Sullivan 2019). It is therefore important to note that different actors with different agendas and political alignments have recommended different forms of participative administration, depending on whether popular participation is considered a means or an end. Often the spaces for discussion created have predefined agendas and, therefore, predefined limits to participation (Turreira-García et al. 2018). As such, one of the challenges to participatory conservation of biodiversity is to create mechanisms that amplify popular participation—as true democracy is measured by the possibility given to people to manifest, including creating new rules and institutions (Safatle 2010).

There appears to be no widely accepted definition of "participatory biodiversity conservation," as the term has been used together with, or as synonyms for, the approaches mentioned above. However, in addition to the effective participation of communities and their institutions in the control and management of natural resources, one aspect is considered central to all of those projects—the conciliation of biodiversity conservation objectives with the desire for benefits, principally economic benefits, to local communities. It is also assumed that participative biodiversity conservation promotes dialogues among all of the players involved, as well as environmental justice and sustainability, and takes into account cultural diversity.

In addition to the observations presented above, other presumptions raised in these chapters must be taken into consideration: (1) The epistemological diversity of the world is potentially infinite, as all social practices involve knowledge. The production of knowledge is therefore, in itself, a social practice. That leads to the conclusion that the universalism of modern science is nothing more than an Occidental mindset, and that there are many other sets of knowledge that are validated by other criteria, not just those of science (Santos 2002). (2) What we call the "environment" is a historically produced social and physical construction. Therefore, what are generally called environmental changes would be better labeled as socio-environmental changes that are never politically neutral, as they invariably produce positive and negative effects for different social groups (Swyngedouw et al. 2002). (3) Biodiversity conservation is not just an act of political and/or technical will, but also a process to be viewed through the lenses of socio-ecological systems and complexity, where structural, institutional, and cultural contexts interact at various levels to generate results (Berkes 2007; Sullivan 2019).

To consider all of the characteristics and presumptions of participatory biodiversity conservation in the current context of the fragility of democracy and threats to human and nonhuman rights is an enormous challenge. It is therefore important to stress that this book does not intend to present a new paradigm for conservation, or a "silver bullet" that could resolve the biodiversity crisis, but rather seeks to advance our search for an inclusive, pluralistic, just and, consequently, more democratic biological and cultural conservation.

Acknowledgments CB thanks the Conselho Nacional de Desenvolvimento Científico e Tecnológico (CNPq) for the Productivity Grant Fellowship (Process number 308628/2016-0).

References

Abessa D, Famá A, Buruaem L (2019) The systematic dismantling of Brazilian environmental laws risks losses on all fronts. Nat Ecol Evol 3:510–511. https://doi.org/10.1038/s41559-019-0855-9

Berkes F (2007) Community-based conservation in a globalized world. Proc Natl Acad Sci U S A 104:15188–15193. https://doi.org/10.1073/pnas.0702098104

Carmo S, Neves R (2019) TCU vai apurar possível "ineficiência" da gestão de Salles no Meio Ambiente. Available via Congresso em Foco. https://congressoemfoco.uol.com.br/meio-ambiente/tcu-vai-apurar-possivel-ineficiencia-da-gestao-de-salles-no-meio-ambiente/. Accessed 22 Jul 2019

Carvalho I (2019) Em 200 dias, Brasil liberou mais agrotóxicos que a União Europeia em oito anos Available via Brasil de Fato. https://www.brasildefato.com.br/2019/07/18/em-200-dias-brasil-liberou-mais-agrotoxicos-que-a-uniao-europeia-em-oito-anos. Accessed 19 Sept 2019.

Correia M (2019) 'Eficiência' do governo para limpar o petróleo do Nordeste: doenças e esgotamento dos voluntários. Available via The intercept Brazil. https://theintercept.com/2019/10/24/voluntarios-intoxicados-oleo-nordeste-bolsonaro/. Accessed 22 Nov 2019

Fellet J (2019) Índios se aliam a antigos inimigos contra planos de Bolsonaro na Amazônia. Available via BBC. https://www.bbc.com/portuguese/brasil-49528317. Accessed 20 Set 2019

Khadka D, Nepal SK (2010) Local responses to participatory conservation in Annapurna Conservation Area, Nepal. Environ Manag 45:351–362. https://doi.org/10.1007/s00267-009-9405-6

Kujawski D (2019) Sínodo da Amazônia: novos caminhos para a igreja e para uma ecologia integral. Available via Pastoral Carcerária. https://carceraria.org.br/igreja-em-saida/sinodo-da-amazonia-novos-caminhos-para-a-igreja-e-para-uma-ecologia-integral. Accessed 01 Dez 2019

Medeiros A, Albarado E (2013) A teoria da ação dialógica no novo cenário da democracia brasileira. In: Colóquio Internacional Paulo Freire VIII. http://coloquio.paulofreire.org.br/participacao/index.php/coloquio/viii-coloquio/paper/view/124/144. Accessed 29 Sept 2019

Mendonça E (2019) Bolsonaro's Brazil unlikely to achieve Paris agreement goals: experts. Available via Mongabay. https://news.mongabay.com/2019/09/bolsonaros-brazil-unlikely-to-achieve-paris-agreement-goals-experts/. Accessed 04 Nov 2019

Monedero J (2012) ¿Posdemocracia? Frente al pesimismo de la nostalgia, el optimismo de la desobediencia. Avaliable via Nueva Sociedad. https://nuso.org/articulo/posdemocracia-frente-al-pesimismo-de-la-nostalgia-el-optimismo-de-la-desobediencia/. Accessed 04 Out 2019

Safatle V (2010) A democracia que não veio. Available via Folha de São Paulo. https://www1.folha.uol.com.br/fsp/poder/po3008201013.htm. Accessed 15 Out 2019

Santos BS (2002) Democratizar a democracia: os caminhos da democracia participativa. Civilização Brasileira, Rio de Janeiro

Schreiber M (2019) STF mantém demarcação de terras indígenas no Ministério da Justiça, ao menos neste ano. Available via BBC. https://www.bbc.com/portuguese/brasil-49187664. Accessed 08 Out 2019

Senra R (2019) Bolsonaro quer revogar decreto ambiental e usar dinheiro saudita para criar 'Cancún brasileira' em Angra. Available via BBC. https://www.bbc.com/portuguese/geral-50229887. Accessed 04 Out 2019

Simões M (2019) Pesquisadora é perseguida após comprovar que não existe dose segura de agrotóxicos. Available via Sul21. https://www.sul21.com.br/ultimas-noticias/geral/2019/09/pesquisadora-e-perseguida-apos-comprovar-que-nao-existe-dose-segura-de-agrotoxicos/. Accessed 12 Nov 2019

Soares I (2019) Decreto de Bolsonaro reduz composição do Conama de 96 conselheiros para 23. Available via Correio Brasiliense Política. https://www.correiobraziliense.com.br/app/noticia/politica/2019/05/29/interna_politica,758531/decreto-de-bolsonaro-reduz-composicao-do-conama-de-100-conselheiros-pa.shtml. Accessed 1 Jul 2019

Strassburg BBN (2019) Conservation provides multiple wins for Brazil. Nat Ecol Evol 3:508–509. https://doi.org/10.1038/s41559-019-0856-8

Sullivan L (2019) Conservation in context: toward a systems framing of decentralized governance and public participation in wildlife management. Rev Policy Res 36:242–261. https://doi.org/10.1111/ropr.12326

Swyngedouw E, Kaika M, Castro E (2002) Urban water: a political-ecology perspective. Built Environ 28:124–137. https://doi.org/10.2307/23288796

Terra de Direitos (2019) In: Ribeiro D, Borges L, Bittencourt N et al. O que está acontecendo no Brasil? Nova gestão presidencial: desmantelamento de políticas sociais e violações de direitos humanos. Available via Terra de Direitos. https://terradedireitos.org.br/uploads/arquivos/NOVO-Informativo-A3%2D%2D-Brasil%2D%2D-4pgs.pdf. Accessed 10 Nov 2019

Trigueiro A (2019) 15 pontos para entender os rumos da desastrosa política ambiental no governo Bolsonaro. Available via G1. https://g1.globo.com/natureza/blog/andre-trigueiro/post/2019/06/03/15-pontos-para-entender-os-rumos-da-desastrosa-politica-ambiental-no-governo-bolsonaro.ghtml. Accessed 12 Out 2019

Turreira-García N, Lund JF, Domínguez P et al (2018) What's in a name? Unpacking "participatory" environmental monitoring. Ecol Soc 23(2):24. https://doi.org/10.5751/ES-10144-230224

Wike R, Silver L, Castillo A (2019) Many across the globe are dissatisfied with how democracy is working. Avaliable via pew research center. https://www.pewresearch.org/global/2019/04/29/many-across-the-globe-are-dissatisfied-with-how-democracy-is-working/. Accessed Set 25 2019

Chapter 2
Multiple Perspectives on Biodiversity Conservation: From Concept to Heated Debate

Cristina Baldauf and Vitor de Oliveira Lunardi

2.1 Biodiversity: One Term, Multiple Interpretations

Many books that address the topic of biological conservation begin with a definition of biodiversity. More than a cliché, this choice is interesting because the term *biodiversity*, besides having become a buzzword, allows several interpretations, which, in turn, influences the ways in which we think about and carry out biological conservation.

The term *biological diversity* was first used in 1968 by the scientist and conservationist, Raymond F. Dasmann, in the book, *Different Kind of Country*, to express the diversity of life forms (Dasmann 1968). However, it was not until the 1980s that this expression, and its contraction (*biodiversity*), became prevalent in science (Soulé and Wilcox 1980; Soulé 1985). One landmark in the consolidation of the use of the word *biodiversity* occurred in 1986 during The National Forum on BioDiversity in Washington, DC, organized by the National Research Council of the National Academy of Sciences and the Smithsonian Institute. The discussions and results of this forum subsidized the publication of the book, *Biodiversity* (National Academy of Sciences 1988), with Edward O. Wilson and Frances M. Peter as editors.

The most widely used definition of biological diversity was proposed by the Convention on Biological Diversity (CBD) during the United Nations Conference on Environment and Development in Rio de Janeiro in 1992: "Biological diversity means the variability among living organisms from all sources including, inter alia, terrestrial, marine and other aquatic ecosystems and the ecological complexes of

C. Baldauf (✉)
Departamento de Biociências, Universidade Federal Rural do Semi-Árido, Mossoró, Rio Grande do Norte, Brazil
e-mail: crisbaldauf@ufersa.edu.br

V. de Oliveira Lunardi
Laboratório de Ecologia Evolutiva e Molecular, Centro de Ciências Biológicas e da Saúde, Universidade Federal Rural do Semi-árido, Mossoró, RN, Brazil

C. Baldauf (ed.), *Participatory Biodiversity Conservation*,
https://doi.org/10.1007/978-3-030-41686-7_2

which they are part; this includes diversity within species, between species and of ecosystems" (CBD 1992). This definition expresses the multidimensionality of the concept, encompassing taxonomic, genetic, and ecological diversity, as well as the ways in which they vary in space and time (see Purvis and Hector 2000; Naeem et al. 2016; Burch-Brown and Archer 2017).

Since the concept of *biological diversity* was launched by the CBD in 1992, *life diversity* and *biodiversity* have become technical terms widely disseminated in newspapers, technical reports, scientific research, books, and political speeches. However, with regard to the use of the word *biodiversity*, inaccuracies are relatively common (see Millennium Ecosystem Assessment 2005 for review). For instance, information on species richness is sometimes used interchangeably with biodiversity, but this often leads to an underestimation because species richness does not include genetic and ecological diversities (Millennium Ecosystem Assessment 2005).

Biodiversity is usually expressed as a unique feature of *pristine* natural ecosystems and/or legally protected natural areas (Millennium Ecosystem Assessment 2005). Therefore, exotic species are frequently excluded from regional estimates of biodiversity. This trend has been questioned since exotic species are an integral component of ecosystems and some may contribute to the maintenance of native species and/or provide important ecosystem services (Schlaepfer 2018). Likewise, exotic species may be an important constituent of what is recognized as *biocultural diversity*, that is, "the diversity of life in all its manifestations: biological, cultural, and linguistic — which are interrelated (and possibly coevolved) within a complex socio-ecological adaptive system" (Maffi et al. 2007). From this perspective, cultural and linguistic diversity are understood as expressions of the evolutionary potential of life (Maffi et al. 2007; Maffi and Dilts 2014) and Indigenous Peoples and Local Communities (IPLC) represent the most significant part of the world's cultural diversity.

From a Political Ecology standpoint, the different views about the term *biodiversity* are not just connected with the multidimensionality of the original concept. Instead, they reveal that, despite the concrete biophysical referents of the term, it can also be seen as a discursive construction that articulates a new relation between nature and society in the global context of science, culture, and economics (Escobar 1998). This *biodiversity discourse*, which is anchored in a narrative of biological crisis, resulted in a network of actors dominated by international institutions, Northern non-governmental organizations (NGOs), botanical gardens, universities, research institutes, and pharmaceutical companies (Escobar 1998).

Biodiversity discourse has also been studied through the lens of Michel Foucault's concept of *biopolitics*. Although its original application was centered on the governance of human life, in Foucauldian terms "biopolitics deals with the population, with the population as a political problem, as a problem that is at once scientific and political, as a biological problem and as power's problem" (Foucault 2003). With regard to biodiversity, a biopolitical approach has been employed to understand how nonhuman individuals and populations are disciplined, aggregated, and optimized in conservation projects, as well as who decides which life should be protected, how, why, and for whom (Biermann and Anderson 2017).

2.2 Biodiversity Loss in the Anthropocene

A multifaceted body of empirical evidence on the influence of humans on other species and on the geophysical world from the late eighteenth century to the present day has led researchers to strongly suggest that this period should be considered as a new epoch in Earth's history—the *Anthropocene* (Crutzen and Stoermer 2000; Steffen et al. 2016; Waters et al. 2016). Among other factors, the elevation of the Earth's average surface temperature above pre-industrial levels—most likely caused by an increase in the concentration of greenhouse gases in the atmosphere (IPCC 2013, 2019)—has been put forward as one of the main reasons for recognizing the Anthropocene as a subdivision of geologic time (see Zalasiewicz et al. 2017).

Those who advocate for the use of the term Anthropocene point out that, in contrast to previous periods of mass extinction in the history of life on Earth, this is the first time that a single highly competitive species (*Homo sapiens*) has been responsible for the extinction of a large number of species (Barnosky et al. 2011; Ceballos et al. 2015). Recent data have revealed that humans have driven at least 680 vertebrate species to extinction since AD 1500, and it is estimated that one million animal and plant species (75%) are currently threatened with extinction (IPBES 2019, but see contrasting estimates [and opinions] in Costello 2019). Even though there is still little agreement among scientists on the numbers and the consequences of today's mass extinction, it is believed that this process will have profound implications for the evolution of biodiversity and human well-being (Ceballos and Ehrlich 2018). Based on our understanding of the current threats to biodiversity, it is estimated that in the near future the rate of species extinctions will surpass even those of the recent past (Johnson et al. 2017).

Reduction in the area of natural habitats, overharvesting, introduction of invasive alien species, pollution, and climate change have been considered the main threats to biodiversity over recent centuries (reviews in Meffe and Carroll 1997; Millennium Ecosystem Assessment 2005; IPBES 2019). In many habitats, some of these threats co-occur, generating synergistic effects that are still poorly understood (Fig. 2.1), but that can generate even more severe negative impacts on biodiversity (Mantyka-pringle et al. 2012). Among the aforementioned threats, the loss of natural habitats is primarily responsible for the loss of biodiversity, since the loss of habitat causes natural populations to lose the resources and the conditions their individuals require to move, reproduce, and sustain life (Meffe and Carroll 1997). Further empirical data from a larger number of species are needed to accurately estimate current rates of extinction and the extent of the impact of habitat loss on natural populations (Rahbek and Colwell 2011; Johnson et al. 2017). However, there is agreement that the loss of natural areas in the tropics has resulted in extensive biodiversity loss due to their high levels of biodiversity prior to habitat conversion/destruction and the often low densities and limited geographic distributions of tropical species (Rahbek and Colwell 2011).

Land-use changes are primarily responsible for the largest relative impact on terrestrial and freshwater ecosystems (IPBES 2019; Fig. 2.2), especially in recent

Fig. 2.1 Deforestation associated with agriculture and extensive cattle ranching in Brazilian Amazon. (**a**) An IBAMA—Brazilian Institute of the Environment and Renewable Natural Resources—operation to combat fires in the Amazon. (**b**) Deforested area for cattle raising in an area adjacent to the Rio Ouro Preto Extractive Reserve, state of Rondônia. (Photos (**a**) Vinícius Mendonça/Ibama; (**b**) Cristina Baldauf)

decades during which a large number of continental natural habitats have been converted into cultivated land (for agriculture, pasture, forestry, and aquaculture) and urban areas, which now cover over one-third of the continental area of the planet (Millennium Ecosystem Assessment 2005; IPBES 2019). Anthropogenic continental areas, which may exhibit low levels of native vegetation cover, degraded native vegetation, low connectivity with other remaining natural areas, and intensively used soil, represent the most likely sites of species extinction and high loss of biodiversity (see Fischer and Lindenmayer 2007). Losses of coastal and marine habitats

Fig. 2.2 Threats to biocultural diversity in Rio Grande do Norte State, northeastern Brazil. (**a**) Invasion of the exotic species *Prosopis juliflora* (Fabaceae). (**b**) Two *Charadrius semipalmatus* (Charadriidae) individuals foraging in a polluted area. (**c**) Salt harvest in mangrove areas impacting both biodiversity and local livelihoods. (Photos (**a**, **b**) Vitor de Oliveira Lunardi; (**c**) Cristina Baldauf)

are less well documented than continental losses, but it is recognized that seabed trawling, reef fishing, and the intensive occupation of coastal areas are some of the major causes of habitat loss in these ecosystems (Millennium Ecosystem Assessment 2005; Airoldi and Beck 2007; Airoldi et al. 2008; Pan et al. 2013).

A detailed analysis of the International Union for Conservation of Nature (IUCN) Red List of Threatened Species published in 2016 revealed that overharvesting is the greatest threat to endangered or near-endangered species (Maxwell et al. 2016). In terrestrial ecosystems, increased demand for timber has led to the overharvesting of many forest species (Sebbenn et al. 2008), while in marine environments, overharvesting of fish for human consumption and for aquaculture has been the main factor in the loss of marine and coastal biodiversity globally (Millennium Ecosystem Assessment 2005; IPBES 2019; see database of global marine fisheries catch 1950–2014 in Watson 2017). In some regions, it is estimated that the biomass of targeted and incidentally caught fish species has been reduced to approximately 90% of that before the advent of industrial sea fishing (Millennium Ecosystem Assessment 2005; Thurstan et al. 2010). Considering that fishing activities currently overexploit fish stocks in many regions, it is unlikely that depleted stocks will recover under a business-as-usual scenario (Neubauer et al. 2013).

The intentional or unintentional introduction of species beyond their natural geographical distributions can lead to biodiversity losses, especially when these species are invasive. Alien invasive species commonly present high reproductive and dispersal rates, genetic diversity, and phenotypic plasticity, as well as being habitat generalists and human commensals (Meffe and Carroll 1997). For example, the introduction of the domestic cat (*Felis domesticus*) in many regions of the planet represents one of the best-known examples of an invasive species' negative impact on continental biodiversity (Medina et al. 2011; Loss et al. 2013; Nogales et al. 2013). In the oceans, maritime transport and aquaculture are the most common means of introductions of invasive alien species, thus threatening marine biodiversity (Molnar et al. 2008). Through analyzing the IUCN Red List, Bellard et al. (2016) were able to identify alien invasive species as the second most common threat associated with species that have become extinct (Bellard et al. 2016; see also IUCN 2019). Current estimates of biological invasions are astounding: "Nearly one fifth of the Earth's surface is at risk of plant and animal invasions, impacting native species, ecosystem functions and nature's contributions to people, as well as economies and human health" (IPBES 2019). The Invasive Species Specialist Group (ISSG) of the IUCN Species Survival Commission (SSC) maintains a free-access online platform for information on invasive alien species that adversely affect biodiversity (http://www.iucngisd.org/gisd/about.php; Fig. 2.2).

Since the 1950s, pollution caused by the pervasive release of synthetic substances into the air, water, and soil, as well as noise and artificial lighting, has been posing major threats to biodiversity. The risks of introducing synthetic substances into ecosystems are of major concern and were first brought to the public's attention in 1962 with the publication of Rachel Carlson's *Silent Spring* (Carlson 2002). Many scientific studies and reports have investigated the effects of synthetic substances on biodiversity, for example: (1) organochlorine, organophosphate, and carbamate cause traumas and serious sublethal effects during reproductive stages in birds (Mitra et al. 2011); (2) the intensive use of nutrient fertilizers, especially nitrogen, phosphorus, and sulfur, has led to biodiversity loss and caused imbalances in the nutrient cycle of terrestrial, freshwater, and coastal ecosystems (Millennium Ecosystem Assessment 2005; Woodward et al. 2012; Vikas and Dwarakish 2015; IPBES 2019); (3) increased atmospheric nitrogen deposition has reduced plant diversity in some natural terrestrial ecosystems (Millennium Ecosystem Assessment 2005; Phoenix et al. 2006); (4) the intense production and use of synthetic plastics have led to the accumulation of this material in large parts of many terrestrial and aquatic ecosystems (Geyer et al. 2017), posing a great threat for some taxonomic groups (e.g., Wilcox et al. 2015; Lamb et al. 2018), "affecting at least 267 species, including 86 per cent of marine turtles, 44 per cent of seabirds and 43 per cent of marine mammals" (IPBES 2019; Fig. 2.2); and (5) the increase in the use of artificial lighting and the production of anthropogenic noises are recently appreciated threats to some animal species (Hölker et al. 2010; Francis and Barber 2013).

Human-driven climate change is projected to be the greatest threat to global biodiversity over the coming decades. Some studies have documented the proximate causes of declining species' abundances resulting from climate change (review in Cahill et al. 2013), while others have focused on changes to species' geographic distributions brought about by shifting climate envelopes (review in Chen et al. 2011). Among Earth's ecosystems, coral reefs have been suffering severely from biodiversity loss due to climate change (Carpenter et al. 2008). The Intergovernmental Panel on Climate Change (IPCC) Special Report on the Ocean and Cryosphere in a Changing Climate provides evidence that the ocean has warmed, become more acidic and less productive, and that extreme coastal events are becoming more severe and sea levels are rising as a result of global warming (IPCC 2019). There are currently many impediments to developing policies and management strategies that effectively contribute to the maintenance of biodiversity, one major challenge being our understanding of, and ability to measure, the synergistic effects of climate change, habitat loss, and other factors (Mantyka-pringle et al. 2012).

It is important to note that all these causes of global biodiversity loss are related to a complex interplay of social, economic, political, and biological factors at different levels (Wood et al. 2000). In this context, there is strong criticism regarding the *globalocentric perspective*—the view of dominant institutions such as the World Bank and several NGOs based in the North—which emphasizes efficient resource management and is based on a particular representation of the threats to biodiversity, rather than focusing on their underlying causes (Escobar 1998). As a response to this *globalocentric* view, several authors have stressed that conventional development, based on the modernity paradigm, have been undermining both biological and cultural diversity in its advocacy for economic growth (Gari 2000; Leff 2004; Brockington and Duffy 2010; Latorre and Latorre 2012; Porto-Gonçalves and Leff 2015).

Finally, despite the broad consensus in the scientific community that the Anthropocene has seen extensive losses of biodiversity, and will continue to do so, a group of scientists have been suggesting that "not all change is bad," so this new era would have the potential to increase biological diversity since populations can evolve, diverge, hybridize, or speciate in "human-made" or *novel* ecosystems (Marris 2011; Thomas 2013). Obviously, such a provocative perspective generates divisions of opinion among the conservation community, a topic that will be addressed in the next session.

2.3 Contrasting Perspectives and Trends in Biodiversity Conservation

The reasons for promoting biodiversity conservation are varied and multifaceted, but can be split into two general schools of philosophical and ethical thought about culture and conservation: on one side are the anthropocentrically oriented social scientists, while the ecocentrically concerned natural scientists and conservation groups reside on the other (Shoreman-Ouimet and Kopnina 2015). Under an anthropocentric view, categories of instrumental values of biodiversity include, but are not limited to, goods (extracted natural resources), services (provided by other species and ecosystems), information (applied science), and psycho-spirituality (aesthetic beauty, religiosity, and sources of scientific knowledge; Meffe and Carroll 1997), which contribute directly and indirectly to human well-being. Loss of biodiversity has resulted in the reduction of this much-needed support system for human well-being, leading, for example, to poverty and reduced access to good quality water (review in Millennium Ecosystem Assessment 2005). Under a non-anthropocentric view, biodiversity represents an ecological and evolutionary heritage, possessing a unique intrinsic value, and this argument alone would justify its conservation (Soulé 1985; Piccolo 2017).

These different perspectives have been present in virtually all debates related to biodiversity conservation; however, the ecocentric view and information from the natural sciences have tended to be what has guided conservation actions (Bennett et al. 2017). Not surprisingly, the creation of protected areas (PAs) is frequently considered the cornerstone of biodiversity conservation strategies (see Dudley 2008; Gray et al. 2016). In 2010, during the 10th Conference of the Parties of the United Nations (COP 10) to the Convention on Biological Diversity (CBD), in Nagoya, 193 signatory countries signed an agreement that stated that "by 2020, at least 17 per cent of terrestrial and inland water areas, and 10 per cent of coastal and marine areas, especially areas of particular importance for biodiversity and ecosystem services, are conserved through effectively and equitably managed, ecologically representative and well connected systems of protected areas and other effective area-based conservation measures, and integrated into the wider landscapes and seascapes" (Aichi Biodiversity Target 11—CBD 2010).

Nearly ten years after COP 10, there have been advances in the creation of new protected areas and currently there are more than 202,000 PAs covering approximately 15% of the Earth's land surface (UNEP-WCMC 2018; see also, Watson et al. 2016; Saura et al. 2018). The oceans have 15,345 PAs, corresponding to 7.44% (26,937,551 km^2) of their total area (UNEP-WCMC and IUCN 2019). However, 32.8% of terrestrial PAs are under intense pressure from human activities (Jones et al. 2018). CBD's Aichi Biodiversity Target 11 was to increase the terrestrial land area covered by well-connected PAs to at least 17% by 2020; this, however, has not been met, with only 7.5% reaching these criteria (Saura et al. 2018).

While the CBD's Aichi Biodiversity Target 11 seems challenging, it is one of the more modest initiatives in terms of conservation goals. For example, the *Nature Needs Half* network aims to protect 50% of Earth's surface by 2030. This initiative received attention recently due to the famous *Half-Earth* campaign, whose central objective is to conserve half of the planet in order to stop the species extinction crisis (Wilson 2016). Proposed by the renowned conservation biologist Edward Wilson, *Half-Earth* is gaining momentum and has attracted a large number of scientists who believe that the only way to stop the biodiversity crisis is to protect as much area as possible from human activities. The proposal to conserve half the planet is hardly novel, as it was suggested by Odum and Odum (1972) more than four decades ago. However, in recent years this idea has not only expanded, but also motivated the development of research evaluating *which* 50% should be conserved to guarantee the protection of biodiversity (see Dinerstein et al. 2017; Pimm et al. 2018).

Whereas many conservationists typically favor strictly protected areas as a primary conservation strategy based on nature's intrinsic value (Soulé 1985, 2014; Doak et al. 2014; Kopnina 2016; Piccolo 2017; Kopnina et al. 2018), the continued loss of biodiversity and its main drivers suggest that even if a significant portion of the Earth's land and water is set aside in PAs, these will function as islands surrounded by profoundly altered landscapes. For this reason, it has been suggested that future conservation efforts should include areas affected by human activities. These suggestions, however, have been criticized by traditional conservationists who blame other conservationists of being "uncaring and prone to be dismissive of the wildlands and the magnificent biodiversity these lands still shelter…" (Wilson 2016, p. 79). In fact, the relative importance of preserved landscapes versus novel ecosystems is at the core of current debates on conservation, such as the land sparing versus land sharing[1] and the *New Conservation Debate*, which was triggered by the articles: *Conservation in the Anthropocene* (Kareiva et al. 2012) and *What is conservation science?* (Kareiva and Marvier 2012). The authors of these articles (the *new conservationists*) claim that Conservation Biology needs a broader framework—which they coined *conservation science*—that incorporates improvement of human well-being as one of its goals. They also argue that, beyond the ethical motivations, nature merits conservation for practical and self-centered reasons (Kareiva and Marvier 2012).

A major criticism of new conservationists lies in their deep engagement with the big corporations (Holmes et al. 2016), particularly with ones that cause worldwide environmental and social impacts, such as timber, mineral, and fossil fuel companies. For instance, CI (Conservation International) has partnered with Chevron, ExxonMobil, Monsanto, Nestlé, and Shell, to name a few, whereas TNC (The Nature Conservancy) has partnered with Cargill, Dow Chemical, PepsiCo, Rio Tinto, among others (Hance 2016). Many of these companies are directly or indirectly involved in deforestation and human rights violations, especially toward traditional and local

[1] The merits of land sparing versus land sharing were originally questioned by Green et al. (2005). Since then, scientists, policy makers, and practitioners have been debating whether agriculture, the main global land-use, should be separated from (land sparing) or integrated within (land sharing) biodiversity-rich areas in order to minimize its ecological impacts. See Perfecto et al. (2009) and Tscharntke et al. (2012) for the advantages of land sharing and the links between agriculture, conservation, and food sovereignty, and Phalan (2018) for a synthesis supporting land sparing. For alternative framings that seek to overcome the dichotomy of the debate around global food system sustainability, see Kremen (2015), Fischer et al. (2017), Wittman et al. (2017), and Vandermeer et al. (2018).

populations (Martínez-Alier 2011), which raises questions about the win–win outcomes for people and nature that new conservationists affirm does happen when working with corporations. On the other hand, engagement with capitalism is rare among traditional conservationists, but one exception can be found in Wilson (2016). The third part of his *Half-Earth* book is entitled *The Solution*, and argues that the combination of technological advancement and free market trade is an efficient mechanism to reduce our collective ecological footprint and solve the biodiversity crisis.

Despite public debate being centered around the dichotomy between biocentric (traditional conservation) and anthropocentric/utilitarian (new conservation) approaches in conservation, other viewpoints exist in this arena. Based on a survey exploring the range of views that exist within the conservation movement, scientists from *The Future of Conservation* project established a typology of conservationists, classifying the respondents into broad types. Besides the *traditional conservation* and *new conservation* perspectives, they also identified two other positions: *market biocentrism*, represented by Wilson's *Half-Earth*, and that which is referred to as *critical social science*, which focuses on improving the well-being of poor populations and is strongly critical of capitalist-based approaches to conservation.

Holmes et al. (2016) conducted another survey of the viewpoints in conservation by sampling the participants of an international conservation conference to assess the range of positions regarding key questions in conservation. Even though they found similar positions as those previously described, their results did reveal a more nuanced view among the interviewees. Some of them supported biocentric approaches, but with less emphasis on wilderness protection than traditional conservationists, whereas others agreed with new conservation perspectives, but with less weight on increasing human well-being as a goal of conservation. These authors also identified another viewpoint—conservation to benefit people, but opposed to capitalist approaches. But they affirm that this viewpoint is almost absent from the published debate on conservation (Holmes et al. 2016), even though this perspective is central to the fields of Political Ecology, Ecological Economics, and in part of the research conducted by defenders of the land-sharing model. In truth, apart from the dramatic confrontation between groups like Greenpeace and capitalist organizations, the conservation movement could be considered moderate in resisting and combating capitalist development (Brockington et al. 2008). This role, however, has been played by several Indigenous Peoples and Local Communities (IPLC) in the defense of their territory, culture, and identity (Escobar 1998, 2015).

The protagonism of IPLC and social movements in the defense of their territories and, consequently, biodiversity is recognized in the conservation typology presented by Escobar (1998). He identified four major positions held by the biodiversity network: (1) *Resource Management*—a globalocentric perspective based on CBD's core objectives and focused on biodiversity threats, but without in-depth discussion of the economic drivers associated with these threats; (2) *Sovereignty*—a common nationalist perspective of developing nations (now "global south") with a view to negotiate the terms of biodiversity treaties and strategies, but without questioning the globalocentric discourse; (3) *Biodemocracy*—a perspective of progressive Southern Hemisphere NGOs that advocate for a shift in attention from south to north as the source of the diversity crisis and for natural resources to be controlled locally, as well as campaigning for the suspension of megadevelopment projects and the recognition of the cultural basis of biological diversity; and (4) *Cultural Autonomy*—a social movements' perspective that has many similarities with that of the Southern Hemisphere NGOs, but with a focus on the defense of their territory, culture, and identity, in other words, the defense of an entire way of life, not just of *resources* or biodiversity.

Escobar's (1998) typology was proposed over two decades ago, but it is still possible to recognize these four positions in the conservation community of today, being evident in international conventions and events related to the environment (e.g., CBD, United Nations Framework Convention on Climate Change, and Rio+20) and their respective side events (Fig. 2.3). For some authors, the overlaps

Fig. 2.3 (**a**) Plenary session at the conference where the current global climate policy framework (The Paris Agreement) was negotiated in December 2015. (**b**) Demonstration entitled "We are dying" held by the Freedom from Debt Coalition, during the climate conference in Lima, Perú, in December 2014. (Photos (**a**, **b**) Jesús Garcia-Latorre)

between positions in conservation have been increasing in recent years, suggesting that the distinctions between conservation perspectives are not as rigid as they used to be (see Mace 2014; Tallis and Lubchenco 2014). On the other hand, Büscher and Fletcher (2019) claim that a complete break with the conservation models conceived so far is essential, since they do not take political and economic realities seriously enough. Thus, they recently proposed a new conservation model called "convivial conservation," which promotes equity, structural transformation, and environmental justice and is based on five key elements: (1) from protected to promoted areas; (2) from saving nature to celebrating human and nonhuman nature, (3) from touristic voyeurism to engaged visitation; (4) from spectacular to everyday environmentalisms, and (5) from natural capital to embedded value(s).

Despite the growing environmental and political crises demanding radical choices, we still can look for commonalities and complementary aspects between viewpoints (Büscher and Fletcher 2019). Therefore, in the next (and final) section, we will discuss the (im)possibilities of dialogue and integration of different positions in biodiversity conservation.

2.4 Final Considerations

The debate over the future of conservation has become less heated over recent years, resulting in the development of new frameworks that attempt to reconcile contrasting viewpoints (Hunter et al. 2014; Chan et al. 2016; Pearson 2016; Gillson et al. 2019). A recent survey with a sample of more than 9000 conservationists revealed that most of them agree on many issues and that their views combine elements of both *people-centered conservation* (new conservation) and *science-led ecocentrism* (traditional conservation; Sambrook et al. 2019).

On the other hand, some authors point out irreconcilable differences among contrasting conservation perspectives. Sambrook et al. (2019) found that the most polarizing questions of their survey included whether it is acceptable to displace people for conservation purposes, a debate that started decades ago and is far from reaching a consensus (see Chap. 4, Sec. 2, "A Brief History of Conservation"). However, the survey also revealed that there is a sense of positivity among the interviewees who believe in win–win solutions for people and nature (Sambrook et al. 2019; Watson and Jones 2019). Concerning the reconciliation of schools of thought on conservation, rather than imposing the opinions of their own field (see Shoreman-Ouimet and Kopnina 2015), conservationists should focus on cooperating to accomplish common goals, avoiding fruitless (and sometimes egocentric) debates that slow progress toward conservation goals (Hunter Jr. et al. 2014; Gavin et al. 2018).

Frameworks focused on overcoming polarized debates, such as conservation based on biocultural approaches (see Gavin et al. 2015), can be more effective at ensuring conservation of biocultural diversity (Gavin et al. 2018; Chap. 11, this volume). However, in view of the incompatibility of socioeconomic development and environmental sustainability (Spaiser et al. 2017), a conservation framework based on structural transformations, such as the convivial conservation, is essential to ensure medium- and long-term biocultural conservation.

Last, but not least, in complex systems such as socio-ecological systems, modern science is only one source of knowledge; thus, partnerships involving Indigenous Peoples and Local Communities, conservation practitioners, and governments are necessary due to their benefits for conservation of ecologically and culturally valuable landscapes (Garnett et al. 2018). This growing trend of democratizing conservation through the involvement of several stakeholders and knowledge sources, despite its challenges, is essential for both biodiversity and cultural conservation to be successful. Examples of meaningful collaborations between researchers and IPLC can be found in the following chapters of this book.

Acknowledgments CB thanks the Conselho Nacional de Desenvolvimento Científico e Tècnológico (CNPq) for the Productivity Grant Fellowship (Process number 308628/2016-0) and the Universidade Federal Rural do Semi-árido (Edital 19/2018) for financial support. VOL received support from Universidade Federal Rural do Semi-árido (Editais 19/2018; 22/2018; 39/2019).

References

Airoldi L, Beck MW (2007) Loss, status and trends for coastal marine habitats of Europe. In: Gibson RN, Atkinson RJA, Gordon JDM (eds) Oceanography and marine biology, annual review, vol 45, 1st edn. CRC Press, Boca Raton, pp 345–405

Airoldi L, Balata D, Beck MW (2008) The gray zone: relationships between habitat loss and marine diversity and their applications in conservation. J Exp Mar Biol Ecol 366(1–2):8–15. https://doi.org/10.1016/j.jembe.2008.07.034

Barnosky AD, Matzke N, Tomiya S et al (2011) Has the Earth's sixth mass extinction already arrived? Nature 471(7336):51–57. https://doi.org/10.1038/nature09678

Bellard C, Cassey P, Blackburn TM (2016) Alien species as a driver of recent extinctions. Biol Lett 12(2):20150623. https://doi.org/10.1098/rsbl.2015.0623

Bennett NJ et al (2017) Mainstreaming the social sciences in conservation. Conserv Biol 31(1):56–66. https://doi.org/10.1111/cobi.12788

Biermann C, Anderson RM (2017) Conservation, biopolitics, and the governance of life and death. Geogr Compass 11(10):e12329. https://doi.org/10.1111/gec3.12329

Brockington D, Duffy R (2010) Capitalism and conservation: the production and reproduction of biodiversity conservation. Antipode 42(3):469–484. https://doi.org/10.1111/j.1467-8330.2010.00760.x

Brockington D, Duffy R, Igoe J (2008) Nature unbound: conservation, capitalism and the future of protected areas. Routledge, London

Burch-Brown J, Archer A (2017) In defence of biodiversity. Biol Philos 32(6):969–997. https://doi.org/10.1007/s10539-017-9587-x

Büscher B, Fletcher R (2019) Towards convivial conservation. Conserv Soc 17(3):283–296. https://doi.org/10.4103/cs.cs_19_75

Cahill AE, Aiello-Lammens ME, Fisher-Reid MC et al (2013) How does climate change cause extinction? Proc R Soc B 280(1750):20121890–20121890. https://doi.org/10.1098/rspb.2012.1890

Carlson R (2002) Silent spring. Houghton Mifflin Harcourt, Boston

Carpenter KE, Abrar M, Aeby G et al (2008) One-third of reef-building corals face elevated extinction risk from climate change and local impacts. Science 321(5888):560–563. https://doi.org/10.1126/science.1159196

CBD (Convention on Biological Diversity) (2010) COP decision X/2. Strategic plan for biodiversity 2011–2020. https://www.cbd.int/decision/cop/?id=12268. Accessed 15 Jan 2019

Ceballos G, Ehrlich PR (2018) The misunderstood sixth mass extinction. Science 360(6393):1080–1081. https://doi.org/10.1126/science.aau0191

Ceballos G, Ehrlich PR, Barnosky AD et al (2015) Accelerated modern human-induced species losses: entering the sixth mass extinction. Sci Adv 1(5):e1400253. https://doi.org/10.1126/sciadv.1400253

Chan KM et al (2016) Opinion: why protect nature? Rethinking values and the environment. Proc Natl Acad Sci U S A 113(6):1462–1465. https://doi.org/10.1073/pnas.1525002113

Chen IC, Hill JK, Ohlemüller R et al (2011) Rapid range shifts of species associated with high levels of climate warming. Science 333(6045):1024–1026. https://doi.org/10.1126/science.1206432

Costello MJ (2019) Unhelpful inflation of threatened species. Science 365(6451):332–333. https://doi.org/10.1126/science.aay3467

Crutzen PJ, Stoermer EF (2000) The "anthropocene". IGBP Glob Chang Newsl 41:17–18

CBD (Convention on Biological Diversity) (1992) The convention on biological diversity. Secretariat of the CBD, U.N. Environment Programme, Montreal. https://www.cbd.int/convention/articles/default.shtml?a=cbd-02. Accessed 01 October 2019

Dasmann RF (1968) A different kind of country. Macmillan, New York. http://agris.fao.org/agris-search/search.do?recordID=US201300589625. Accessed 19 Jan 2019

Dinerstein E, Olson D, Joshi A (2017) An ecoregion-based approach to protecting half the terrestrial realm. Bioscience 67(6):534–545. https://doi.org/10.1093/biosci/bix014

Doak DF, Bakker VJ, Goldstein BE, Hale B. (2014). What is the future of conservation? Trends in Ecolgy & Evolution 29(2): 77–81

Dudley N (ed) (2008) Guidelines for applying protected area management categories. IUCN, Gland. https://portals.iucn.org/library/node/30018. Accessed 2 Jan 2019

Escobar A (1998) Whose knowledge, whose nature? Biodiversity, conservation, and the political ecology of social movements. J Polit Ecol 5(1):53–82. https://doi.org/10.2458/v5i1.21397

Escobar A (2015) Territorios de diferencia: la ontología política de los "derechos al territorio". Cuadernos de antropología social 41:25–38. ISSN: 0327-3776. Available at: https://www.redalyc.org/articulo.oa?id=180942587002

Fischer J, Lindenmayer DB (2007) Landscape modification and habitat fragmentation: a synthesis. Glob Ecol Biogeogr 16(3):265–280. https://doi.org/10.1111/j.1466-8238.2007.00287.x

Fischer J, Abson DJ, Bergsten A et al (2017) Reframing the food–biodiversity challenge. Trends Ecol Evol 32(5):335–345. https://doi.org/10.1016/j.tree.2017.02.009

Foucault M (2003) Society must be defended: lectures at the Collège de France 1975–1976. English edition: (2003) (trans. David Macey). Picador: New York, 242

Francis CD, Barber JR (2013) A framework for understanding noise impacts on wildlife: an urgent conservation priority. Front Ecol Environ 11(6):305–313. https://doi.org/10.1890/120183

Gari JA (2000) The political ecology of biodiversity: biodiversity conservation and rural development at the indigenous and peasant grassroots. Thesis Ph.D. University of Oxford

Garnett ST, Burgess ND, Fa JE, Fernández-Llamazares Á et al (2018) A spatial overview of the global importance of indigenous lands for conservation. Nat Sustainability 1(7):369–374. https://doi.org/10.1038/s41893-018-0100-6

Gavin MC, McCarter J, Mead A et al (2015) Defining biocultural approaches to conservation. Trends Ecol Evol 30(3):140–145. https://doi.org/10.1016/j.tree.2014.12.005

Gavin MC et al (2018) Effective biodiversity conservation requires dynamic, pluralistic, partnership-based approaches. Sustainability 10(6):1846. https://doi.org/10.3390/su10061846

Geyer R, Jambeck JR, Law KL (2017) Production, use, and fate of all plastics ever made. Sci Adv 3(7):e1700782. https://doi.org/10.1126/sciadv.1700782

Gillson L, Biggs H, Smit IP et al (2019) Finding common ground between adaptive management and evidence-based approaches to biodiversity conservation. Trends Ecol Evol 34(1):31–44. https://doi.org/10.1016/j.tree.2018.10.003
Gray CL, Hill SLL, Newbold T (2016) Local biodiversity is higher inside than outside terrestrial protected areas worldwide. Nat Commun 7:12306. https://doi.org/10.1038/ncomms12306
Green RE, Cornell SJ, Scharlemann JPW et al (2005) Farming and the fate of wild nature. Science 307:550–555. https://doi.org/10.1126/science.1106049
Hance J (2016) How big donors and corporations shape conservation goals. Mongabay series: evolving conservation. Available via DIALOG. https://news.mongabay.com/2016/05/big-donors-corporations-shape-conservation-goals/. Accessed 15 out 2019
Hölker F, Wolter C, Perkin EK et al (2010) Light pollution as a biodiversity threat. Trends Ecol Evol 25(12):681–682. https://doi.org/10.1016/j.tree.2010.09.007
Holmes G, Sandbrook C, Fisher JA (2016) Understanding conservationists' perspectives on the new-conservation debate. Conserv Biol 31(2):353–363. https://doi.org/10.1111/cobi.12811
Hunter JR, Malcolm L, Redford KH et al (2014) The complementary niches of anthropocentric and biocentric conservationists. Conserv Biol 28(3):641–645. https://doi.org/10.1111/cobi.12296
IPBES (2019) In: Díaz S, Settele J, Brondizio ES et al (eds) Summary for policymakers of the global assessment report on biodiversity and ecosystem services of the Intergovernmental Science-Policy Platform on Biodiversity and Ecosystem Services. IPBES Secretariat, Bonn, Germany. https://www.ipbes.net/global-assessment-report-biodiversity-ecosystem-services. Accessed 20 Sep 2019
IPCC (2013) In: Stocker TF, Qin D, Plattner G-K et al (eds) Climate change 2013: the physical science basis. Contribution of Working Group I to the Fifth assessment report of the intergovernmental panel on climate change. Cambridge University Press, Cambridge. https://www.ipcc.ch/report/ar5/wg1/. Accessed 02 Jan 2019
IPCC (2019) Summary for policymakers. In: Pörtner H-O, Roberts DC, Masson-Delmotte V, Zhai P, Tignor M, Poloczanska E, Mintenbeck K, Nicolai M, Okem A, Petzold J, Rama B, Weyer N (eds) IPCC Special Report on the ocean and cryosphere in a changing climate. In press. https://www.ipcc.ch/srocc/. Accessed 15 Nov 2019
IUCN (2019) Invasive species. https://www.iucn.org/theme/species/our-work/invasive-species. Accessed 25 Jan 2019
Johnson CN, Balmford A, Brook BW et al (2017) Biodiversity losses and conservation responses in the Anthropocene. Science 356(6335):270–275. https://doi.org/10.1126/science.aam9317
Jones KR, Venter O, Fuller RA et al (2018) One-third of global protected land is under intense human pressure. Science 360(6390):788–791. https://doi.org/10.1126/science.aap9565
Kareiva P, Marvier M (2012) What is conservation science? Bioscience 62(11):962–969. https://doi.org/10.1525/bio.2012.62.11.5
Kareiva P, Marvier M, Lalasz R (2012) Conservation in the anthropocene: beyond solitude and fragility. The Break Through Institute. Available via DIALOG. https://thebreakthrough.org/journal/issue-2/conservation-in-the-anthropocene. Accessed 10 set 2019
Kopnina H (2016) Half the earth for people (or more)? Addressing ethical questions in conservation. Biol Conserv 203:176–185. https://doi.org/10.1016/j.biocon.2016.09.019
Kopnina H, Washington H, Gray J et al (2018) The 'future of conservation' debate: defending ecocentrism and the nature needs half movement. Biol Conserv 217:140–148. https://doi.org/10.1016/j.biocon.2017.10.016
Kremen C (2015) Reframing the land-sparing/land-sharing debate for biodiversity conservation. Ann N Y Acad Sci 1355(1):52–76. https://doi.org/10.1111/nyas.12845
Lamb JB, Willis BL, Fiorenza EA et al (2018) Plastic waste associated with disease on coral reefs. Science 359(6374):460–462. https://doi.org/10.1126/science.aar3320
Latorre JG, Latorre JG (2012) Globalization, local communities, and traditional forest-related knowledge. In: Traditional forest-related knowledge. Springer, Dordrecht, pp 449–490
Leff E (2004) Racionalidad ambiental: la reapropiación social de la naturaleza. Siglo XXI, Buenos Aires, Argentina

Loss SR, Will T, Marra PP (2013) The impact of free-ranging domestic cats on wildlife of the United States. Nat Commun 4:1396. https://doi.org/10.1038/ncomms2380

Mace GM (2014) Whose conservation? Science 345(6204):1558–1560. https://doi.org/10.1126/science.1254704

Maffi L, Dilts O (2014). Biocultural diversity toolkit, vol 1: Introduction to biocultural diversity. Terralingua, Salt Spring Island

Maffi L et al (2007) Biocultural diversity and sustainability. In: The SAGE handbook of environment and society. SLE Pound, London, pp 267–277

Mantyka-pringle CS, Martin TG, Rhodes JR (2012) Interactions between climate and habitat loss effects on biodiversity: a systematic review and meta-analysis. Glob Chang Biol 18(4): 1239–1252. https://doi.org/10.1111/j.1365-2486.2011.02593.x

Marris ERG (2011) Saving nature in a post-wild world. Bloomsbury, New York

Martínez-Alier J (ed) (2011) El ecologismo de los pobres: conflictos ambientales y lenguajes de valoración, 5rd edn. Icaria, Barcelona, p 416

Maxwell SL, Fuller RA, Brooks TM et al (2016) Biodiversity: the ravages of guns, nets and bulldozers. Nature 536(7615):143–145. https://doi.org/10.1038/536143a

Medina FM, Bonnaud E, Vidal E et al (2011) A global review of the impacts of invasive cats on island endangered vertebrates. Glob Chang Biol 17(11):3503–3510. https://doi.org/10.1111/j.1365-2486.2011.02464.x

Meffe GK, Carroll CR (eds) (1997) Principles of conservation biology. Sinauer Associates, Sunderland

Millennium Ecosystem Assessment (2005) Ecosystems and human well-being: biodiversity synthesis. World Resources Institute, Washington, DC

Mitra A, Chatterjee C, Mandal FB (2011) Synthetic chemical pesticides and their effects on birds. Res J Environ Toxicol 5(2):81–96. https://doi.org/10.3923/rjet.2011.81.96

Molnar JL, Gamboa RL, Revenga C et al (2008) Assessing the global threat of invasive species to marine biodiversity. Front Ecol Environ 6(9):485–492. https://doi.org/10.1890/070064

Naeem S, Prager C, Weeks B et al (2016) Biodiversity as a multidimensional construct: a review, framework and case study of herbivory's impact on plant biodiversity. P R Soc B 283(1844):20153005. https://doi.org/10.1098/rspb.2015.3005

National Academy of Sciences (1988) Biodiversity. The National Academies Press, Washington, DC. https://doi.org/10.17226/989

Neubauer P, Jensen OP, Hutchings JA et al (2013) Resilience and recovery of overexploited marine populations. Science 340(6130):347–349. https://doi.org/10.1126/science.1230441

Nogales M, Vidal E, Medina FM et al (2013) Feral cats and biodiversity conservation: the urgent prioritization of island management. Bioscience 63(10):804–810. https://doi.org/10.1525/bio.2013.63.10.7

Odum EP, Odum HT (1972) Natural areas as necessary components of man's total environment. Trans North Am Wildl Nat Res Conf 37:178–189

Pan J, Marcoval MA, Bazzini SM et al (2013) Coastal marine biodiversity challenges and threats. In: Arias AH, Menendez MC (eds) Marine ecology in a changing world, 1st edn. CRC Press, New York, pp 43–67

Pearson RG (2016) Reasons to conserve nature. Trends Ecol Evol 31(5):366–371. https://doi.org/10.1016/j.tree.2016.02.005

Perfecto I, Vandermeer J, Wright A (2009) Nature's matrix: linking agriculture, conservation and food sovereignty. Routledge, London

Phalan B (2018) What have we learned from the land sparing-sharing model? Sustainability 10(6):1760. https://doi.org/10.3390/su10061760

Phoenix GK, Hicks WK, Cinderby S et al (2006) Atmospheric nitrogen deposition in world biodiversity hotspots: the need for a greater global perspective in assessing N deposition impacts. Glob Chang Biol 12(3):470–476. https://doi.org/10.1111/j.1365-2486.2006.01104.x

Piccolo JJ (2017) Intrinsic values in nature: objective good or simply half of an unhelpful dichotomy? J Nat Conserv 37:8–11. https://doi.org/10.1016/j.jnc.2017.02.007

Pimm SL, Jenkins CN, Li BV (2018) How to protect half of Earth to ensure it protects sufficient biodiversity. Sci Adv 4(8):eaat2616. https://doi.org/10.1126/sciadv.aat2616

Porto-Gonçalves CW, Leff E (2015) Political ecology in Latin America: the social re-appropriation of nature, the reinvention of territories and the construction of an environmental rationality. Desenv e Meio Ambiente 35(1):65–88. https://doi.org/10.5380/dma.v35i0.43543

Purvis A, Hector A (2000) Getting the measure of biodiversity. Nature 405(6783):212–219. https://doi.org/10.1038/35012221

Rahbek C, Colwell RK (2011) Species loss revisited. Nature 473(7347): 288–289. https://doi.org/10.1038/473288a

Sambrook C, Fisher JA, Holmes G et al (2019) The global conservation movement is diverse but not divided. Nat Sustainability 2(4):316–323. https://doi.org/10.1038/s41893-019-0267-5

Saura S, Bertzky B, Bastin L et al (2018) Protected area connectivity: shortfalls in global targets and country-level priorities. Biol Conserv 219:53–67. https://doi.org/10.1016/j.biocon.2017.12.020

Schlaepfer MA (2018) Do non-native species contribute to biodiversity? PLoS Biol 16(4):e2005568. https://doi.org/10.1371/journal.pbio.2005568

Sebbenn AM, Degen B, Azevedo VC et al (2008) Modelling the long-term impacts of selective logging on genetic diversity and demographic structure of four tropical tree species in the Amazon forest. For Ecol Manage 254(2):335–349. https://doi.org/10.1016/j.foreco.2007.08.009

Shoreman-Ouimet E, Kopnina H (2015) Reconciling ecological and social justice to promote biodiversity conservation. Biol Conserv 184:320–326. https://doi.org/10.1016/j.biocon.2015.01.030

Soulé ME (1985) What is conservation biology? A new synthetic discipline addresses the dynamics and problems of perturbed species, communities, and ecosystems. Bioscience 35(11):727–734. https://doi.org/10.2307/1310054

Soulé M (2014) The "new conservation". In: Wuerthner G, Crist E, Butler T (eds) Keeping the wild. Island Press, Washington, DC, pp 66–80. https://doi.org/10.5822/978-1-61091-559-5_7

Soulé ME, Wilcox BA (eds) (1980) Conservation biology: an evolutionary- ecological perspective. Sinauer, Sunderland

Spaiser V, Ranganathan S, Swain RB, Sumpter, DJ (2017) The sustainable development oxymoron: quantifying and modelling the incompatibility of sustainable development goals. Int J Sust Dev World Ecol 24(6):457–470. https://doi.org/10.1080/13504509.2016.1235624

Steffen W, Leinfelder R, Zalasiewicz J et al (2016) Stratigraphic and Earth System approaches to defining the Anthropocene. Earth's Future 4(8):324–345. https://doi.org/10.1002/2016ef000379

Tallis H, Lubchenco J (2014) Working together: a call for inclusive conservation. Nat News 515(7525):27–28. https://doi.org/10.1038/515027a

Thomas CD (2013) The Anthropocene could raise biological diversity. Nature News 502(7469):7–7. https://doi.org/10.1038/502007a

Thurstan RH, Brockington S, Roberts CM (2010) The effects of 118 years of industrial fishing on UK bottom trawl fisheries. Nat Commun 1:15. https://doi.org/10.1038/ncomms1013

Tscharntke T et al (2012) Global food security, biodiversity conservation and the future of agricultural intensification. Biol Conserv 151(1):53–59. https://doi.org/10.1016/j.biocon.2012.01.068

UNEP-WCMC (2018) Protected areas map of the world, April 2018. http://www.protectedplanet.net. Accessed 02 Feb 2019

UNEP-WCMC, IUCN (2019) Marine protected planet, January 2019. http://www.protectedplanet.net. Accessed 01 Feb 2019

Vandermeer J et al (2018) Feeding prometheus: an interdisciplinary approach for solving the global food crisis. Front Sustain Food Syst 2:39. https://doi.org/10.3389/fsufs.2018.00039

Vikas M, Dwarakish GS (2015) Coastal pollution: a review. Aquat Procedia 4:381–388. https://doi.org/10.1016/j.aqpro.2015.02.051

Waters CN, Zalasiewicz J, Summerhayes C et al (2016) The Anthropocene is functionally and stratigraphically distinct from the Holocene. Science 351(6269):aad2622. https://doi.org/10.1126/science.aad2622

Watson R (2017) A database of global marine commercial, small-scale, illegal and unreported fisheries catch 1950–2014. Sci Data 4:170039. https://doi.org/10.1038/sdata.2017.39

Watson JEM, Jones JPG (2019) Diverse contributions benefit people and nature. Nat Ecol Evol 3:1140–1141. https://doi.org/10.1038/s41559-019-0936-9
Watson JE, Darling ES, Venter O et al (2016) Bolder science needed now for protected areas. Conserv Biol 30(2):243–248. https://doi.org/10.1111/cobi.12645
Wilcox C, Van Sebille E, Hardesty BD (2015) Threat of plastic pollution to seabirds is global, pervasive, and increasing. Proc Natl Acad Sci U S A 112(38):11899–11904. https://doi.org/10.1073/pnas.1502108112
Wilson EO (2016) Half-earth: our planet's fight for life. WW Norton & Company, New York
Wittman H et al (2017) A social–ecological perspective on harmonizing food security and biodiversity conservation. Reg Environ Chang 17(5):1291–1301. https://doi.org/10.1007/s10113-016-1045-9
Wood A, Stedman-Edwards P, Mang J (2000) The root causes of biodiversity loss. World Wildlife Fund and Earthscan Publications, Ltd, London, p 304
Woodward G, Gessner MO, Giller PS et al (2012) Continental-scale effects of nutrient pollution on stream ecosystem functioning. Science 336(6087):1438–1440. https://doi.org/10.1126/science.1219534
Zalasiewicz J, Waters CN, Wolfe AP et al (2017) Making the case for a formal Anthropocene Epoch: an analysis of ongoing critiques. Newsl Stratigr 50(2):205–226. https://doi.org/10.1127/nos/2017/0385

Part II
Beyond "Fortress Conservation": Reflections and Experiences

Chapter 3
Domesticated Nature: The Culturally Constructed Niche of Humanity

Charles R. Clement, Carolina Levis, Juliano Franco-Moraes, and André Braga Junqueira

3.1 Introduction

The myth of the pristine (Denevan 1992) and that of wilderness (Cronon 1996) are based on the idea that today there are areas that represent the world in its primitive state, prior to human intervention. These myths are creations of natural historians and environmentalists of the nineteenth and twentieth centuries, and fascinate conservationists and many natural scientists, whose models for understanding Nature often exclude humans (Ellis and Ramankutty 2008). Both myths were built on the Nature-Culture dichotomy in Western academies (Descola 2013; Glacken 1967), and are based on the premise that humans are a threat to Nature, so that areas without visible human intervention must be kept separate from humans because these "wilderness" areas represent Nature in its pure state (Diegues 2008). This "threat" is the basis of the Anthropocene (Crutzen 2002), a geological epoch in which global

C. R. Clement (✉)
Instituto Nacional de Pesquisas da Amazônia, Manaus, AM, Brazil
e-mail: cclement@inpa.gov.br

C. Levis
Programa de Pós-Graduação em Ecologia, Instituto Nacional de Pesquisas da Amazônia, Manaus, AM, Brazil

Forest Ecology and Forest Management Group, Wageningen University & Research, Wageningen, The Netherlands

J. Franco-Moraes
Department of Ecology, Institute of Biosciences, Universidade de São Paulo, São Paulo, SP, Brazil

A. B. Junqueira
International Institute for Sustainability, Rio de Janeiro, RJ, Brazil

Institut de Ciència i Tecnologia Ambientals, Universitat Autònoma de Barcelona, Barcelona, Spain

C. Baldauf (ed.), *Participatory Biodiversity Conservation*,
https://doi.org/10.1007/978-3-030-41686-7_3

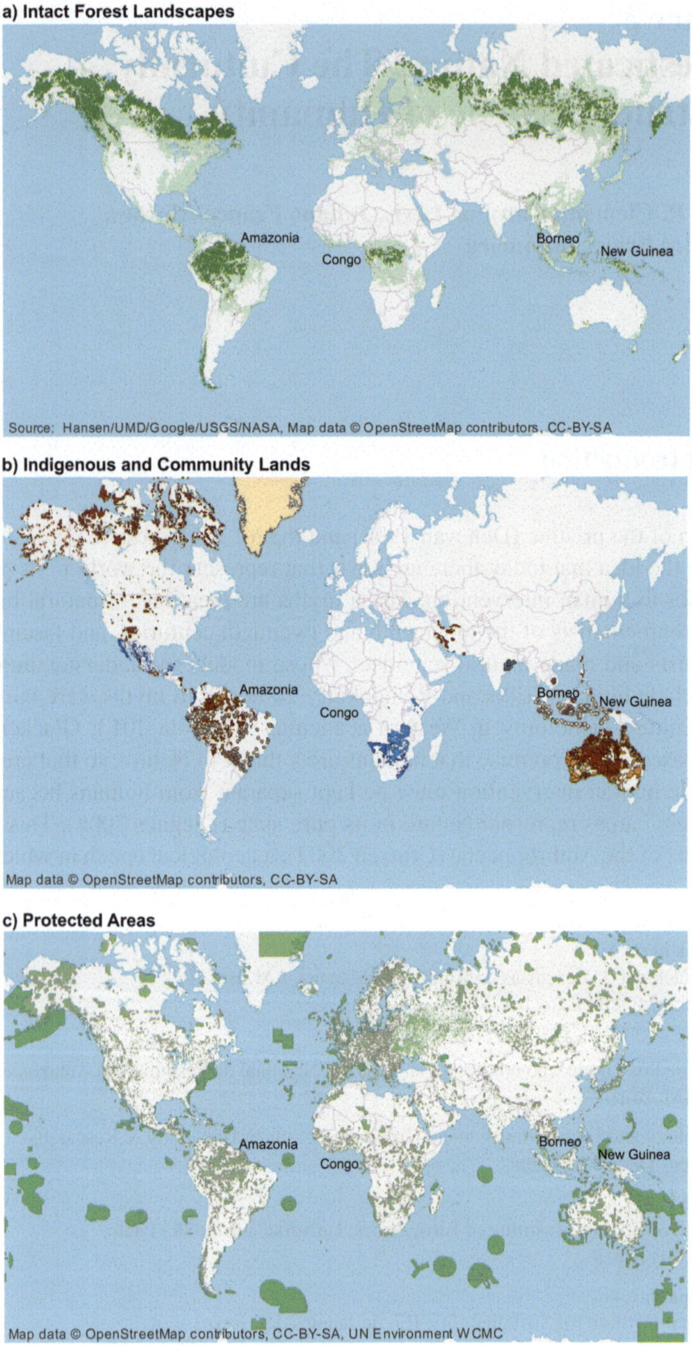

Fig. 3.1 Global distributions of forest landscapes, indigenous territories and local community lands, and protected areas. (**a**). Forest landscapes, defined as areas with >30% tree cover at the turn

industrial society's degradation of the biosphere is challenging human survival (Ripple et al. 2017). The fact that these myths continue to influence conservation proposals is worrying, in that basing conservation on myths is likely to lead to outcomes that are far from those that are hoped for (Büscher et al. 2016; Vandermeer and Perfecto 2014). As an example, Edward O. Wilson's recent Half-Earth proposal (2016) envisions a network of interconnected protected areas, mostly free of human intervention, covering half of the planet, to save biodiversity from humanity. However, this vision ignores the rapidly growing body of theory and evidence that shows that our species has constructed large parts of our niches across the planet since the late Pleistocene (Laland et al. 2001), so much so that we have become the ultimate ecosystem engineers (Ellis 2011) and can talk about domesticated Nature (Kareiva et al. 2007), because Nature in its pure state does not exist anymore (Cronon 1996; Denevan 1992; Diegues 2008). It also ignores numerous examples where conservation works best when people are involved, not only as funders and enthusiasts, but as actors in landscapes that they have domesticated (Büscher et al. 2016; Linnell et al. 2015; Oldekop et al. 2016). The wilderness × domesticated debate is current, even though some conservationists recognize the importance of people in conservation (Büscher et al. 2016; Watson et al. 2018). Here we outline evidence from the last century and the current millennium to show that Nature was domesticated well before the advent of the Anthropocene in 1750 AD (Crutzen 2002).

The planet has approximately 13 billion hectares (bi ha) of land surface area (Tilman 2012), of which 4 bi ha (31%) are in the Arctic and Antarctic, or are desert or tundra, with low primary productivity and very-low-density human populations. The other 9 bi ha (69%) are domesticated and therefore interest most humans. The catchall term "agriculture," which includes both farming and animal husbandry by smallholders and agrobusinesses, occupies about 5 bi ha (38%), with 75% of this in pastures. Silviculture and managed forests occupy another 1.5 bi ha (11%). That leaves about 2.5 bi ha (9%), which include built environments, but mostly forests, woodlands, and woody savannas mixed with some natural grasslands that are relatively less "impacted by development" (the part of this area with >20% tree cover in forest landscapes is presented in Fig. 3.1a). These areas are considered to have small human footprints (Jacobson et al. 2019; Pimm et al. 2018; Venter et al. 2016) and are called "wilderness," "intact forest landscapes," or "low impact areas" (Jacobson et al. 2019; Potapov et al. 2017; Watson et al. 2018), even though people live in most of them. The use of this terminology ignores the fact that words have histories and power, which in this case can evoke the myths we mention above.

(**Fig. 3.1** continued) of the millennium (2000), from Hansen et al. (2013) in light green, and areas considered to be "intact" in dark green (with >20% tree cover), because they are minimally influenced by modern global industrial society, from Potapov et al. (2017). (**b**). Indigenous territories are defined as "the collectively-held and governed lands (and natural resources) of Indigenous Peoples" and Local community lands as "lands that fall under the customary governance of the community whether or not this is recognized in national law"; the bolder the colors, the stronger the legal recognition of indigenous and community rights over their lands. (**c**). Recognized protected areas, both national and international. (Definitions and maps (**b**). and (**c**). from LandMark (2017))

These 2.5 bi ha that are less impacted by global industrial society include designated indigenous territories and local community lands (Fig. 3.1b), designated conservation areas (Fig. 3.1c), private lands, and areas with unclear title that are claimed by national governments, but which are also generally occupied by indigenous peoples and local communities who consider it theirs also. When Kareiva et al. (2007) wrote about domesticated Nature, they focused primarily on the 50% of the global terrestrial surface that is serving global industrial society directly, although they mention that people live in the 19% with less development. Here, we focus exactly on these 19%, because indigenous people and local communities domesticated these terrestrial ecosystems in ways that are less visible to most members of global society. These 19% are also prime real estate for the Half-Earth proponents, especially in the tropics (Pimm et al. 2018), so it is extremely important to show that not only are people in these areas, but they are co-responsible for the Nature that is found there and are key actors in its conservation.

3.2 Niche Construction and Landscape Domestication

Niche construction theory postulates that all organisms modify the biotic/abiotic components of their environments, and that this has consequences for their own evolution (Odling-Smee et al. 2003). Think of birds building nests, ants creating colonies, and earthworms modifying the soil. Some organisms modify their environments so significantly that they are called ecosystem engineers (Jones et al. 1994), such as elephants, alligators, and termites. The most dramatic example of these ecosystem engineers is our own species, *Homo sapiens* (Ellis 2011). By the beginning of the current millennium, the niche construction perspective had gained visibility in ecology and evolution (Odling-Smee et al. 2003).

Since humans are the most potent ecosystem engineers, it is appropriate to speak of cultural niche construction (Laland et al. 2001). Humans accumulate and transmit knowledge, innovations, and technology between generations in numerous ways according to their cultures (Boyd et al. 2011). This social transmission affects both human and cultural evolution (O'Brien and Laland 2012). Two of the most important facets of cultural niche construction, which are closely inter-related, are the domestication of populations of plants and animals (Meyer and Purugganan 2013; Smith 2007, 2011), and the domestication of landscapes (Clement and Cassino 2018). These activities are particularly important because they increase the carrying capacity of landscapes (Laland and O'Brien 2010), although sometimes landscape domestication can also lead to degradation (Diamond 2005; Scott 2017).

Let us first look at the definition of domestication itself. The Oxford English Dictionary definition of the term is (1) To make, or settle as, a member of a household; to cause to be at home; to naturalize; (2) To make to be or to feel "at home"; to familiarize. When Darwin (1859) used domestication as a metaphor to present his ideas about evolution, he emphasized selection by humans, both conscious and, especially, unconscious. He observed that humans select and then accumulate; in other words, they bring their selections home so that they can become familiar. In traditional communities around the world, and even in rural England of the nineteenth century, the home is not just a house, but its associated gardens, fields for crops and animals, agroforests, woodlots, and nearby forests (Scott 2017). With time, these different parts of the landscape, subjected to different types and intensities of management, are gradually converted into domesticated landscapes with greater human carrying capacity.

As mentioned above, while landscape domestication involves increasing the carrying capacity of landscapes, it may also result in degradation, depending on the stance from which this process is viewed and its outcomes analyzed. To "degrade" is "lower the character or quality of" (The Oxford English Dictionary). A soybean field in southern Amazonia is a domesticated landscape, in which food provision for the international market is increased compared to the native ecosystem that it replaced. This increase in crop productivity, however, occurred at the expense of native biodiversity (Soares-Filho et al. 2006) and other important ecosystem services, such as water and nutrient cycling (Lovejoy and Nobre 2018). As Diamond (2005) pointed out, some production systems can be pushed from a productive state to a degraded state through overexploitation, lack of knowledge or investment, which can even lead to the collapse of the whole socio-ecological system. Soybean fields and managed forests are both domesticated landscapes, although very different from each other: the former a product of modern industrial society, with little biodiversity, and totally dependent on machinery and chemicals to be perpetuated; the latter a product of indigenous peoples and local communities, with abundant biodiversity and dependent on traditional human knowledge and management to be maintained. The understanding of gardens, fields (also called swiddens in some systems), agroforests, and woodlots as domesticated landscapes is relatively straightforward. What about the forests managed by locals?

Levis et al. (2018) recently showed how local communities and indigenous people domesticate Amazonian forests, often seen as pristine Nature (Denevan 1992). Levis et al. described how Amazonian people have manipulated—consciously or not—the distribution and abundance of useful plants in the forests around their communities. These actions were classified into eight categories of management practices: (1) removal of non-useful plants, (2) protection of useful plants, (3) attraction of non-human animal dispersers, (4) dispersal of useful plants, (5) selection of phenotypes, (6) fire management, (7) planting of useful plants, and (8) soil improvement. Categories 4, 5, and 7 are straight out of Darwin's ideas about domestication. Along with categories 6 and 8, people manage their gardens, fields, agroforests, and woodlots also. In many parts of the world, fallows are important parts of food production systems, since gardens and swiddens are often fallowed to restore soil fertility (Kleinman et al. 1995), resulting in second-growth forests that can mature if allowed to (Chazdon 2014). In indigenous food production systems, trees and annual crops are cultivated in gardens, swiddens and fallows, and agroforests. When these systems become mature forests, annual crops disappear, but trees persist (Clement 1999). Although these forests do not appear to be domesticated to the inexperienced eye, a careful analysis of their species composition reveals high richness and abundance of perennial species with managed or domesticated populations, a clear legacy of both domestications (Levis et al. 2012, 2017, 2018). Management categories 1, 2, and 3 are practiced primarily in standing forests, especially along trails used by people gathering non-timber forest products and by hunters. Over time, as trails shift within the forest, patches of useful species expand and new patches are initiated, creating a mosaic of patches with more or less human influence. These management practices are not labor intensive and some even occur unconsciously, for example, as people discard seeds of fruits they munch on while walking along the trail, but when they return to the same trail they often protect new seedlings of these fruit trees (Levis et al. 2018; Posey 1985). Hence, domesticating a forest requires only two things: people who practice some of these niche construction activities (categories 1–8) and time. The result is a forest with small and large patches domesticated to different degrees and dominated by useful species favored by these long-term human activities. Throughout the Holocene, people around the world have been doing this in their local forests (Roberts 2019; Roberts et al. 2017).

As might be expected, there is a continuum in the intensity of management and in the degree of landscape domestication surrounding human settlements, ranging from intensively managed gardens, fields, swiddens, and fallows to less intensive agroforestry and woodlots to even less intensively managed forests. This gradient of landscape domestication is associated with a continuum of plant population domestication, ranging from more domesticated crops in gardens and nearby fields and swiddens, to semi-domesticated crops in agroforests and incipiently domesticated crops in managed forests (see Clement (1999) for full descriptions of these domestication continua). All these different types of domesticated landscapes also contain wild species (some of which are useful, some not), which tend to be more diverse and abundant in less intensively managed landscapes. The persistence of these wild species in domesticated landscapes is evidence of Nature's agency within our anthropocentric telling of landscape domestication; this is why we say that humans are co-responsible for the Nature in their landscapes, since these landscapes contain domesticates as well as other biodiversity.

In all of the landscapes that we described as domesticated, wild plants, animals, and microorganisms are present. Viewed anthropocentrically, they live in our niche; seen from their viewpoint, we live in their niches, since every organism in any ecosystem lives in its own niche while also living in the niches of other organisms (Pulliam 2000). So, when we say that a garden, field, agroforest, or forest is domesticated, that is our viewpoint, but thousands of other organisms live there as well. These organisms represent Nature and each has agency, with which they construct parts of their own niches and affect us. If we look at the continua within the landscape mosaic mentioned above, we can see an inverse relationship between human agency and Nature's agency: closer to the home (or community), human agency is more visible to humans than Nature's agency; as one moves away from the home, human agency diminishes and Nature's agency increases. This complexity of agencies and their interactions is at the core of concepts such as coupled human and natural systems (Liu et al. 2007) or socio-ecological systems (Ostrom 2009). It follows that Nature has been domesticated by humans to a greater or lesser extent across the planet and each continent became domesticated once human societies expanded (Ellis and Ramankutty 2008). The "intact forest landscapes" of Potapov et al. (2017) are mostly domesticated by indigenous peoples and local communities (Fig. 3.1a, b), even where indigenous and local community rights are not recognized by national governments.

3.3 Domesticated Nature

Here, we concentrate on a part of the 50% of the terrestrial biosphere identified by Pimm et al. (2018) that is already included in protected areas or has potential for being included in the network of interconnected protected areas envisioned by Wilson (2016), especially in the tropics. Within the 50% are the 19% that we identified earlier as forests, woodlands, and savannas with high levels of biodiversity, which includes various types of protected areas and others with no protection (Fig. 3.1). Much of these 19% have been little impacted by global industrial society and have been called "intact forest ecosystems" (Potapov et al. 2017). Pimm et al. (2018) used a low threshold in a quantitative estimate of the human footprint (as calculated by Venter et al. (2016)) to define areas with low human impact and called them "wilderness", bringing new dimensions and definitions to the term defined by Cronon (1996).

In order to estimate the global human footprint, Venter et al. (2016) used eight variables that are typical of modern global industrial society: built environments, croplands, pasture lands, population density, nightlights, railways, major roadways, and navigable waterways. Built environments and crop and pasture lands were discussed above as domesticated landscapes and are typical of the landscape domestication practiced by global industrial society, indigenous peoples, and local communities. Traditional societies today live in areas with low population densities and few nightlights or navigable waterways, and no railways or roadways, although these may occasionally cut across their lands. The maps generated by Venter et al. (2016) show large areas with low human footprints in Amazonia, the Congo, New Guinea, and Borneo, all of which are strong candidates for inclusion in the Half-Earth network (Pimm et al. 2018) and are subject to rapid and large-scale land use change (see below). One of these four areas just highlighted (Borneo) is a previously declared biodiversity hotspot (Myers et al. 2000) and the other three are high biodiversity regions in general (Amazonia, the Congo, and New Guinea) (Mittermeier et al. 2003) (Fig. 3.1).

Interestingly, all of these areas have high linguistic diversity (Gorenflo et al. 2012). In other words, these prime areas for future conservation are the home of indigenous peoples, with thousands of distinct languages, many of which are already threatened with extinction (Gorenflo et al. 2012; Loh and Harmon 2014; Sutherland 2003), and each of these peoples has its own traditional ecological knowledge about niche construction (Boyd et al. 2011). Although the existence, rights, and voices of these populations are recognized in some scientific discourses (Büscher et al. 2016; Linnell et al. 2015; Oldekop et al. 2016; Vandermeer and Perfecto 2014; Watson et al. 2018), this is not always emphasized in recent conservation proposals (Pimm et al. 2018; Wilson 2016) and even less so by national governments and policies (Vandermeer and Perfecto 2014). Each of these areas will be briefly summarized here, but we emphasize that all of the hotspots and the other regions of high biodiversity are also the home of indigenous peoples and local communities who are co-responsible for their Nature.

Borneo: The earliest unequivocal archaeological evidence of human landscape domestication dates to about 45,000 years ago (Roberts 2019; Roberts et al. 2017). The island is also a very early center of crop domestication (Meyer et al. 2012). Domestication of Borneo's tropical forests has been extensively reported (Michon 2005; Roberts 2019; Roberts et al. 2017; Wiersum 1997, 2004), with a suite of niche construction practices (Wiersum 1997). These long-term sustainable agroforestry systems are overlooked by decision-makers in the national government that divide the island; instead, Indonesia and Malaysia encourage large-scale clearing for oil palm and other industrial agricultural systems (Vijay et al. 2016). Indonesia is building new infrastructure in the area to support this expansion of industrial agriculture (Alamgir et al. 2019). As a consequence, many indigenous peoples are pushed off of their lands and their languages are threatened and at risk of extinction (Gorenflo et al. 2012), along with their biodiversity (Myers et al. 2000). Some indigenous territories are recognized on Borneo, but many fewer than the number of languages (Fig. 3.1b).

New Guinea: The earliest unequivocal archaeological evidence of human landscape domestication dates to about 45,000 years ago (Roberts 2019; Roberts et al. 2017). Like Borneo, New Guinea is a very early center of crop domestication (Meyer et al. 2012). Domestication of highland New Guinea forested landscapes during the early and mid-Holocene has been intensively studied (Denham et al. 2016; Golson 2016; Roberts 2019), and Holocene history highlights the importance of domesticated forests (Kennedy 2012). As in Indonesia, the Government of Papua New Guinea is encouraging forest clearance for oil palm (Vijay et al. 2016). Although less severe in New Guinea, at least for now, the expansion of large-scale agroindustry is pushing peoples off of their lands and increasing the risks of language and biodiversity extinction (Gorenflo et al. 2012). New Guinea has the highest linguistic diversity in the world, with 976 endemic languages, 889 of which have fewer than 10,000 speakers and are thus classified as vulnerable (Gorenflo et al. 2012), and more than 50% of New Guinea's languages are threatened with extinction (Loh and Harmon 2014). The Papua New Guinea government increasingly ignores these people in its development plans (Vijay et al. 2016), although the Indonesian government recognizes some indigenous territories (Fig. 3.1b).

The Congo: The search for archaeological evidence for human landscape domestications in Africa has focused more on wooded savannas than on tropical rainforests (Roberts 2019; Roberts et al. 2017), but there is growing evidence that African rainforests were modified considerably (Morin-Rivat et al. 2017; Van Gemerden et al. 2003). Unlike Borneo and New Guinea, there is not much evidence for crop domestication in the Congo. This may be because only recently have researchers started to look at what may be incipiently domesticated tree crops (Dawson et al. 2014). Some pigmy groups manage yams (*Dioscorea*) in forest patches (Yasuoka 2013), as do people on Borneo and New Guinea. As elsewhere, the various national governments that include parts of the Congo in their jurisdictions are encouraging agricultural expansion to support population growth (Potapov et al. 2012; Tyukavina et al. 2018) and this is expected to expand significantly in the coming decades (Molotoks et al. 2018), so language diversity and biodiversity are also threatened, and languages showed declines of 20–30% in the last decades (Gorenflo et al. 2012). This language extinction is favored by the total lack of recognition of indigenous territories by governments with jurisdiction over parts of the Congo (Fig. 3.1b).

Amazonia: As in the other high-biodiversity areas, language diversity and biodiversity are correlated (Gorenflo et al. 2012). Humans started managing forest resources (Roosevelt et al. 1996) and domesticating plants since the beginning of the Holocene (Clement et al. 2010). In southwestern Amazonia, Watling et al. (2018) recently identified early-to-mid-Holocene cultivation of beans (*Phaseolus* spp) and squash (*Cucurbita* spp), probable management of local forests with guava (*Psidium* spp) and piquiá (*Caryocar* spp), and mid-Holocene cultivation of maize (*Zea mays*), which originated in Mexico. In the same region, a species of rice (*Oryza* spp) was domesticated in the mid-Holocene (Hilbert et al. 2017), and further west cacao (*Theobroma cacao*) started to be domesticated at the same time (Zarrillo et al. 2018). These crops and many others were distributed throughout South America and as far as Mexico (Clement et al. 2010).

Unlike Borneo, New Guinea, and the Congo, which suffered from European conquest and colonization, Amazonia's native peoples were decimated by European conquest and colonization, with its diseases, slavery, and warfare (Denevan 2014; Mann 2005). Indigenous populations collapsed by 90–95% in the centuries immediately following European conquest (Denevan 2014). By the time Europe's first natural historians started to explore the region in the late eighteenth century, these populations had disappeared and the forest had covered their tracks (Mann 2005). This was the origin of the pristine myth in Amazonia, elsewhere in the Americas, Africa, Asia, and Oceania (Denevan 1992). European colonial administrations throughout the tropics contributed to this, because it was convenient to displace people from their forests and guarantee administrative control to facilitate European ideas of development (Fairhead and Leach 2014; Morrison 2014; Vandermeer and Perfecto 2014). In Amazonia, the work of well-known US archaeologist Betty Meggers contributed to support the idea of pristine forests: environmental determinism, which resulted in her idea of a "counterfeit paradise" (Meggers 1996). The idea of environmental determinism starts from the observation that the majority of Amazonian soils are nutrient poor (particularly in uplands and in central and eastern Amazonia), and proceeds to affirm that food production is limited by soils and climatic oscillations, since only small-scale shifting cultivation was observed among the indigenous peoples. With the expansion of archaeological, paleo-ecological, and ecological research in Amazonia during the last few decades, a new understanding of human impacts emerged in the region, now incorporating the niche construction perspective. Archaeologists have identified earthworks, anthropogenic soils, pre-conquest settlement sites that all supported large populations across Amazonia (Heckenberger and Neves 2009). More recently, ecologists have been incorporating archaeological datasets to examine the relationships between forest composition and proximity to archaeological sites, and identified different degrees of forest domestication across the region (Levis et al. 2017). Taken together, this evidence supports the idea that Amazonia is a domesticated biome (Clement et al. 2015), similar to all other biomes across the planet where human populations were and are abundant (Ellis 2011; Ellis and Ramankutty 2008).

In the early twentieth century, the rubber boom first led to the reoccupation of forests emptied by the decimation of indigenous peoples, and then collapsed during the second decade of the century (Hecht 2013). The surviving rubber tappers became fur traders for several decades, over-hunting forests and rivers (Antunes et al. 2016), and only in mid-century did the Brazilian government start to invest in policies to integrate Amazonia into the rest of the country (Souza 2009). During the military regime (1964–1985), these policies caused an arc of deforestation starting from the central Brazilian savannas (the *Cerrado*) and expanding northward that continued during the succeeding democratic governments until today (Nobre et al. 2016). During the last third of the twentieth century, most other Amazonian countries developed similar policies, although with less investment (Souza 2009). As elsewhere, the expansion of global industrial society into Amazonian forests is causing biodiversity extinction (Nobre et al. 2016) and threatening indigenous peoples (Loh and Harmon 2014). Despite important advances in the last decades in reducing deforestation rates, and in the recognition and demarcation of indigenous territories and protected areas (Fig. 3.1b, c), threats to biodiversity, indigenous rights, and language diversity persist and are increasing due to policy changes by the recently elected government in Brazil (Fearnside 2018).

What is clear from this very short review of these four tropical forest areas of high biodiversity is that they are also areas of high cultural diversity, whose diverse human populations expanded their cultural niches during millennia. It follows that what Venter et al. (2016), Potapov et al. (2017), and Pimm et al. (2018) call wilderness, wildlands, or intact forest landscapes are in reality landscapes domesticated to some degree by indigenous societies. So why do so many conservationists ignore this human co-responsibility for Nature?

3.4 From Hotspots to Hope Spots

One of the reasons is the modern society's fascination with the pristine (Denevan 1992) and wilderness (Cronon 1996) myths; as we have seen, leading conservationists continue to use the terminology "wilderness" when discussing areas occupied by indigenous peoples and local communities (e.g., Pimm et al. 2018; Venter et al. 2016; Watson et al. 2018). Another reason is that European conquest and colonization, followed by liberation movements that had adopted many European ideas, caused amnesia about what had existed previously (Morrison 2014; Roberts 2019).

Often, post-liberation national governments purposely avoided rediscovering their histories, because this facilitated state control over areas occupied by "backward" populations (Fairhead and Leach 2014; Vandermeer and Perfecto 2014). With state control comes encouragement of modern development, supported by international development banks/agencies, with their industrial style development that generally starts by removing indigenous populations and local communities, and clearing the forest for "productive" activities (Vandermeer and Perfecto 2014). When international conservation agencies and NGOs complain, national governments sometimes create conservation areas (Fig. 3.1b, c), frequently removing indigenous peoples and local communities, or limiting their options for landscape domestication activities. In short, modern global industrial society creates hotspots and also a few conservation areas to salve our conscious. What is perfectly evident also is that this timid effort at conservation is insufficient, in terms of biodiversity (e.g., Wilson 2016), in terms of social justice (Hecht 2014), and in terms of cultural diversity (Gorenflo et al. 2012; Loh and Harmon 2014).

Although we have stated here that many conservationists are fascinated by the pristine and wilderness myths, there is a growing realization that conservation must be about people as well, especially the people who are co-responsible for the Nature in their territories. There are numerous currents within the conservation movement that advocate for an increased engagement of indigenous people in conservation initiatives and decisions (Mace 2014). Biocultural approaches, for example, suggest that conservation should sustain both biodiversity and cultural diversity, respecting and incorporating the different worldviews and knowledge/management systems of indigenous peoples and local communities (Gavin et al. 2015). Associated with this, another recent current includes the new Intergovernmental Platform on Biodiversity and Ecosystem Services (IPBES), where a large number of participants support the inclusion of indigenous peoples and local communities in conservation planning, monitoring, and territorial management (Díaz et al. 2018; IPBES 2019). They also support the inclusion of traditional ecological knowledge as a valid source of information about biodiversity and landscapes (Díaz et al. 2015; IPBES 2019). Although these new ideas have created controversy in the conservation community (Masood 2018), they are stimulating new efforts around the world, as will be seen in numerous chapters of this volume.

At the recent Belém +30 meeting in 2018 of the International Society for Ethnobiology and the Sociedade Brasileira de Etnobiologia e Etnoecologia, anthropologist Michael Heckenberger argued that it is time to shift from hotspots to hope spots. Hope spots were proposed by Sylvia Earle in 2009 as special places that are vital to the health of the world's oceans (Earle 2016). This idea is also being used to re-think the conservation of terrestrial ecosystems (Rezende et al. 2018; Scarano and Ceotto 2015). These authors argue that participatory approaches for ecosystem conservation and restoration can reduce vulnerability to climate change and improve human welfare. Heckenberger's proposal adds a new idea to expand the implementation of terrestrial hope spots by focusing both on hotspots and other regions where there is associated cultural diversity, precisely the areas of exceptional biocultural diversity (Gorenflo et al. 2012; Loh and Harmon 2014). However, to achieve this, conservation planning needs to incorporate not only environmental sciences but also social sciences (Bennett et al. 2017). If the world conservation community were to adopt this interdisciplinarity, shift its focus to hope spots, and create alliances with indigenous peoples and local communities, the very first requirement is to adopt participatory biodiversity conservation in the spirit of the IPBES and this volume.

Acknowledgments We thank Susanna B. Hecht, UCLA, Berkley, for critical review and useful suggestions.

References

Alamgir M, Campbell MJ, Sloan S, Suhardiman A, Supriatna J, Laurance WF (2019) High-risk infrastructure projects pose imminent threats to forests in Indonesian Borneo. Sci Rep 9(1):140. https://doi.org/10.1038/s41598-018-36594-8

Antunes AP, Fewster RM, Venticinque EM, Peres CA, Levi T, Rohe F, Shepard GH Jr (2016) Empty forest or empty rivers? A century of commercial hunting in Amazonia. Sci Adv 2(10):e1600936. https://doi.org/10.1126/sciadv.1600936

Bennett NJ, Roth R, Klain SC, Chan K, Christie P, Clark DA, Cullman G, Curran D, Durbin TJ, Epstein G, Greenberg A, Nelson MP, Sandlos J, Stedman R, Teel TL, Thomas R, Veríssimo D, Wyborn C (2017) Conservation social science: understanding and integrating human dimensions to improve conservation. Biol Conserv 205:93–108. https://doi.org/10.1016/j.biocon.2016.10.006

Boyd R, Richerson PJ, Henrich J (2011) The cultural niche: why social learning is essential for human adaptation. Proc Natl Acad Sci U S A 108(Suppl 2):10918–10925. https://doi.org/10.1073/pnas.1100290108

Büscher B, Fletcher R, Brockington D, Sandbrook C, Adams WM, Campbell L, Corson C, Dressler W, Duffy R, Gray N, Holmes G, Kelly A, Lunstrum E, Ramutsindela M, Shanker K (2016) Half-earth or whole earth? Radical ideas for conservation, and their implications. Oryx 51(03):407–410. https://doi.org/10.1017/s0030605316001228

Chazdon RL (2014) Second growth: the promise of tropical forest regeneration in an age of deforestation. University of Chicago Press, Chicago

Clement CR (1999) 1492 and the loss of Amazonian crop genetic resources. I. The relation between domestication and human population decline. Econ Bot 53(2):188–202. https://doi.org/10.1007/BF02866498

Clement CR, Cassino MF (2018) Landscape domestication and archaeology. In: Smith C (ed) Encyclopedia of global archaeology. Springer, New York, pp 1–8. https://doi.org/10.1007/978-3-319-51726-1_817-2

Clement CR, Cristo-Araújo M, Coppens d'Eeckenbrugge G, Alves Pereira A, Picanço-Rodrigues D (2010) Origin and domestication of native Amazonian crops. Diversity 2(1):72–106. https://doi.org/10.3390/d2010072

Clement CR, Denevan WM, Heckenberger MJ, Junqueira AB, Neves EG, Teixeira WG, Woods WI (2015) The domestication of Amazonia before European conquest. Proc R Soc B 282(1812):20150813. https://doi.org/10.1098/rspb.2015.0813

Cronon W (1996) The trouble with wilderness: or, getting back to the wrong Nature. Environ Hist 1(1):7–28. https://doi.org/10.2307/3985059

Crutzen PJ (2002) Geology of mankind. Nature 415(6867):23. https://doi.org/10.1038/415023a

Darwin C (1859) On the origin of species by means of natural selection, or the preservation of the favoured races in the struggle for life. John Murray, London

Dawson IK, Leakey R, Clement CR, Weber JC, Cornelius JP, Roshetko JM, Vinceti B, Kalinganire A, Tchoundjeu Z, Masters E, Jamnadass R (2014) The management of tree genetic resources and the livelihoods of rural communities in the tropics: non-timber forest products, smallholder agroforestry practices and tree commodity crops. For Ecol Manage 333:9–21. https://doi.org/10.1016/j.foreco.2014.01.021
Denevan WM (1992) The pristine myth: the landscape of the Americas in 1492. Ann Assoc Amer Geogr 82(3):369–385. https://doi.org/10.1111/j.1467-8306.1992.tb01965.x
Denevan WM (2014) Estimating Amazonian Indian numbers in 1492. J Latin Amer Geogr 13(2):203–217
Denham TP, Iriarte J, Vrydaghs L (2016) Early to mid-Holocene plant exploitation in New Guinea: towards a contingent interpretation of agriculture. In: Denham TP, Iriarte J, Vrydaghs L (eds) Rethinking agriculture: archaeological and ethnoarchaeological perspectives. Routledge, New York, pp 86–116
Descola P (2013) Beyond nature and culture (trans: Lloyd J). University of Chicago Press, Chicago
Diamond J (2005) Collapse: how societies choose to fail or succeed. Penguin, New York
Díaz S, Demissew S, Joly C, Lonsdale WM, Larigauderie A (2015) A Rosetta Stone for nature's benefits to people. PLoS Biol 13(1):e1002040. https://doi.org/10.1371/journal.pbio.1002040
Díaz S, Pascual U, Stenseke M, Martín-López B, Watson RT, Molnár Z, Hill R, Chan KMA, Baste IA, Brauman KA (2018) Assessing nature's contributions to people. Science 359(6373):270–272. https://doi.org/10.1126/science.aap8826
Diegues AC (2008) O mito da natureza intocada, 6th edn. Editora Hucitec, São Paulo
Earle S (2016) Protect the ocean, protect ourselves. In: Johnson M, Bayley A (eds) Coastal change, ocean conservation and resilient communities. Springer, Cham, pp 155–161. https://doi.org/10.1007/978-3-319-41914-5_11
Ellis EC (2011) Anthropogenic transformation of the terrestrial biosphere. Phil Trans R Soc A 369(1938):1010–1035. https://doi.org/10.1098/rsta.2010.0331
Ellis EC, Ramankutty N (2008) Putting people in the map: anthropogenic biomes of the world. Front Ecol Environ 6(8):439–447. https://doi.org/10.1890/070062
Fairhead J, Leach M (2014) False forest history, complicit social analysis: rethinking some West African environmental narratives. In: Hecht SB, Morrison KD, Padoch C (eds) The social lives of forests: past, present, and future of woodland resurgence. University of Chicago Press, Chicago, 4–30
Fearnside PM (2018) Why Brazil's new president poses an unprecedented threat to the Amazon. Yale Environ 360, pp 1–10 https://e360.yale.edu/features/why-brazils-new-president-poses-an-unprecedented-threat-to-the-amazon
Gavin MC, McCarter J, Mead A, Berkes F, Stepp JR, Peterson D, Tang R (2015) Defining biocultural approaches to conservation. Trends Ecol Evol 30(3):140–145. https://doi.org/10.1016/j.tree.2014.12.005
Glacken CJ (1967) Traces on the Rhodian shore: nature and culture in Western thought from ancient times to the end of the eighteenth century. University of California Press, Berkley
Golson J (2016) Unravelling the story of early plant exploitation in Highland Papua New Guinea. In: Denham TP, Iriarte J, Vrydaghs L (eds) Rethinking agriculture: archaeological and ethnoarchaeological perspectives. Routledge, New York, pp 117–133
Gorenflo LJ, Romaine S, Mittermeier RA, Walker-Painemilla K (2012) Co-occurrence of linguistic and biological diversity in biodiversity hotspots and high biodiversity wilderness areas. Proc Natl Acad Sci U S A 109(21):8032–8037. https://doi.org/10.1073/pnas.1117511109
Hansen MC, Potapov PV, Moore R, Hancher M, Turubanova SA, Tyukavina A, Thau D, Stehman SV, Goetz SJ, Loveland TR, Kommareddy A, Egorov A, Chini L, Justice CO, Townshend JR (2013) High-resolution global maps of 21st-century forest cover change. Science 342(6160):850–853. https://doi.org/10.1126/science.1244693
Hecht SB (2013) The scramble for the Amazon and the "Lost paradise" of Euclides da Cunha. University of Chicago Press, Chicago
Hecht SB (2014) Rethinking social lives and forest transitions: history, ideologies, institutions, and the matrix. In: Hecht SB, Morrison KD, Padoch C (eds) The social lives of forests: past, present, and future of woodland resurgence. University of Chicago Press, Chicago, pp 11–13

Heckenberger MJ, Neves EG (2009) Amazonian archaeology. Ann Rev Anthropol 38:251–266. https://doi.org/10.1146/annurev-anthro-091908-164310

Hilbert L, Neves EG, Pugliese F, Whitney BS, Shock M, Veasey E, Zimpel CA, Iriarte J (2017) Evidence for mid-Holocene rice domestication in the Americas. Nat Ecol Evol 1(11):1693–1698. https://doi.org/10.1038/s41559-017-0322-4

IPBES (2019) Global assessment report on biodiversity and ecosystem services of the Intergovernmental Science-Policy Platform on Biodiversity and Ecosystem Services. IPBES Secretariat, Bonn, Germany

Jacobson AP, Riggio J, Tait AM, Baillie JEM (2019) Global areas of low human impact ('Low impact areas') and fragmentation of the natural world. Sci Rep 9(1):14179. https://doi.org/10.1038/s41598-019-50558-6

Jones CG, Lawton JH, Shachak M (1994) Organisms as ecosystem engineers. Oikos 69(3):373–386. https://doi.org/10.2307/3545850

Kareiva P, Watts S, McDonald R, Boucher T (2007) Domesticated nature: shaping landscapes and ecosystems for human welfare. Science 316(5833):1866–1869. https://doi.org/10.1126/science.1140170

Kennedy J (2012) Agricultural systems in the tropical forest: a critique framed by tree crops of Papua New Guinea. Quat Int 249:140–150. https://doi.org/10.1016/j.quaint.2011.06.020

Kleinman PJA, Pimentel D, Bryant RB (1995) The ecological sustainability of slash-and-burn agriculture. Agric Ecosyst Environ 52(2–3):235–249. https://doi.org/10.1016/0167-8809(94)00531-I

Laland KN, O'Brien MJ (2010) Niche construction theory and archaeology. J Arch Method Theory 17(4):303–322. https://doi.org/10.1007/s10816-010-9096-6

Laland KN, Odling-Smee J, Feldman MW (2001) Cultural niche construction and human evolution. J Evol Biol 14(1):22–33. https://doi.org/10.1046/j.1420-9101.2001.00262.x

LandMark: the global platform of indigenous and community lands (2017) World Resources Institute. http://www.landmarkmap.org/

Levis C, Souza PF, Schietti J, Emilio T, da Veiga Pinto JLP, Clement CR, Costa FRC (2012) Historical human footprint on modern tree species composition in the Purus-Madeira interfluve, central Amazonia. PLoS One 7(11):e48559. https://doi.org/10.1371/journal.pone.0048559

Levis C, Costa FRC, Bongers F, Peña-Claros M, Clement CR, Junqueira AB, Neves EG, Tamanaha EK, Figueiredo FOG, Salomão RP, Castilho CV, Magnusson WE, Phillips OL, Guevara JE, Sabatier D, Molino J-F, López DC, Mendoza AM, Pitman NCA, Duque A, Vargas PN, Zartman CE, Vasquez R, Andrade A, Camargo JL et al (2017) Persistent effects of pre-Columbian plant domestication on Amazonian forest composition. Science 355(6328):925–931. https://doi.org/10.1126/science.aal0157

Levis C, Flores BM, Moreira PA, Luize BG, Alves RP, Franco-Moraes J, Lins J, Konings E, Peña-Claros M, Bongers F, Costa FRC, Clement CR (2018) How people domesticated Amazonian forests. Front Ecol Evol 5:171. https://doi.org/10.3389/fevo.2017.00171

Linnell JDC, Kaczensky P, Wotschikowsky U, Lescureux N, Boitani L (2015) Framing the relationship between people and nature in the context of European conservation. Conserv Biol 29(4):978–985. https://doi.org/10.1111/cobi.12534

Liu J, Dietz T, Carpenter SR, Alberti M, Folke C, Moran E, Pell AN, Deadman P, Kratz T, Lubchenco J, Ostrom E, Ouyang Z, Provencher W, Redman CL, Schneider SH, Taylor WW (2007) Complexity of coupled human and natural systems. Science 317(5844):1513–1516. https://doi.org/10.1126/science.1144004

Loh J, Harmon D (2014) Biocultural diversity: threatened species, endangered languages, vol 1. WWF Netherlands, Zeist

Lovejoy TE, Nobre CA (2018) Amazon tipping point. Sci Adv 4:eaat2340. https://doi.org/10.1126/sciadv.aat2340

Mace GM (2014) Whose conservation? Science 345(6204):1558–1560. https://doi.org/10.1126/science.1254704

Mann CC (2005) 1491: new revelations of the Americas before Columbus. Alfred A. Knopf, New York
Masood E (2018) The battle for the soul of biodiversity. Nature 560:423–425. https://doi.org/10.1038/d41586-018-05984-3
Meggers BJ (1996) Amazonia: man and culture in a counterfeit paradise, revised edition, 2nd edn. Smithsonian Institution Press, Washington, DC
Meyer RS, Purugganan MD (2013) Evolution of crop species: genetics of domestication and diversification. Nat Rev Genet 14(12):840–852. https://doi.org/10.1038/nrg3605
Meyer RS, DuVal AE, Jensen HR (2012) Patterns and processes in crop domestication: an historical review and quantitative analysis of 203 global food crops. New Phytol 196(1):29–48. https://doi.org/10.1111/j.1469-8137.2012.04253.x
Michon G (2005) Domesticating forests: how farmers manage forest resources. Institut de Recherche pour le Développement (IRD), Centre for International Forestry Research (CIFOR), The World Agroforestry Centre (ICRAF), Bogor
Mittermeier RA, Mittermeier CG, Brooks TM, Pilgrim JD, Konstant WR, Fonseca GAB, Kormos C (2003) Wilderness and biodiversity conservation. Proc Natl Acad Sci U S A 100(18):10309–10313. https://doi.org/10.1073/pnas.1732458100
Molotoks A, Stehfest E, Doelman J, Albanito F, Fitton N, Dawson TP, Smith P (2018) Global projections of future cropland expansion to 2050 and direct impacts on biodiversity and carbon storage. Glob Chang Biol 24:5895–5908. https://doi.org/10.1111/gcb.14459
Morin-Rivat J, Fayolle A, Favier C, Bremond L, Gourlet-Fleury S, Bayol N, Lejeune P, Beeckman H, Doucet J-L (2017) Present-day central African forest is a legacy of the 19th century human history. Elife 6:e20343. https://doi.org/10.7554/eLife.20343
Morrison KD (2014) Human-forest relationships and the erasure of history. In: Hecht SB, Morrison KD, Padoch C (eds) The social lives of forests: past, present, and future of woodland resurgence. University of Chicago Press, Chicago, pp 143–147
Myers N, Mittermeier RA, Mittermeier CG, Fonseca GAB, Kent J (2000) Biodiversity hotspots for conservation priorities. Nature 403(6772):853–858. https://doi.org/10.1038/35002501
Nobre CA, Sampaio G, Borma LS, Castilla-Rubio JC, Silva JS, Cardoso M (2016) Land-use and climate change risks in the Amazon and the need of a novel sustainable development paradigm. Proc Natl Acad Sci U S A 113(39):10759–10768. https://doi.org/10.1073/pnas.1605516113
O'Brien MJ, Laland KN (2012) Genes, culture, and agriculture. Curr Anthropol 53(4):434–470. https://doi.org/10.1086/666585
Odling-Smee FJ, Laland KN, Feldman MW (2003) Niche construction: the neglected process in evolution. Princeton University Press, Princeton
Oldekop JA, Holmes G, Harris WE, Evans KL (2016) A global assessment of the social and conservation outcomes of protected areas. Conserv Biol 30(1):133–141. https://doi.org/10.1111/cobi.12568
Ostrom E (2009) A general framework for analyzing sustainability of social-ecological systems. Science 325(5939):419–422. https://doi.org/10.1126/science.1172133
Pimm SL, Jenkins CN, Li BV (2018) How to protect half of Earth to ensure it protects sufficient biodiversity. Sci Adv 4(8):eaat2616. https://doi.org/10.1126/sciadv.aat2616
Posey DA (1985) Indigenous management of tropical forest ecosystems: the case of the Kayapo Indians of the Brazilian Amazon. Agr Syst 3(2):139–158. https://doi.org/10.1007/BF00122640
Potapov PV, Turubanova SA, Hansen MC, Adusei B, Broich M, Altstatt A, Mane L, Justice CO (2012) Quantifying forest cover loss in Democratic Republic of the Congo, 2000–2010, with Landsat ETM+ data. Remote Sens Environ 122:106–116. https://doi.org/10.1016/j.rse.2011.08.027
Potapov P, Hansen MC, Laestadius L, Turubanova S, Yaroshenko A, Thies C, Smith W, Zhuravleva I, Komarova A, Minnemeyer S (2017) The last frontiers of wilderness: tracking loss of intact forest landscapes from 2000 to 2013. Sci Adv 3(1):e1600821. https://doi.org/10.1126/sciadv.1600821

Pulliam HR (2000) On the relationship between niche and distribution. Ecol Lett 3(4):349–361. https://doi.org/10.1046/j.1461-0248.2000.00143.x

Rezende CL, Scarano FR, Assad ED, Joly CA, Metzger JP, Strassburg BBN, Tabarelli M, Fonseca GA, Mittermeier RA (2018) From hotspot to hopespot: an opportunity for the Brazilian Atlantic Forest. Perspec Ecol Conserv 16(4):208–214. https://doi.org/10.1016/j.pecon.2018.10.002

Ripple WJ, Wolf C, Newsome TM, Galetti M, Alamgir M, Crist E, Mahmoud MI, Laurance WF, 15,364 scientist signatories from 184 countries (2017) World scientists' warning to humanity: a second notice. Bioscience 67(12):1026–1028. https://doi.org/10.1093/biosci/bix125

Roberts P, Hunt C, Arroyo-Kalin M, Evans D, Boivin N (2017) The deep human prehistory of global tropical forests and its relevance for modern conservation. Nat Plants 3(8):17093. https://doi.org/10.1038/nplants.2017.93

Roberts P (2019) Tropical forests in prehistory, history, and modernity. Oxford University Press, Oxford, UK

Roosevelt AC, Lima da Costa M, Lopes Machado C, Michab M, Mercier N, Valladas H, Feathers J, Barnett W, Imazio da Silveira M, Henderson A, Sliva J, Chernoff B, Reese DS, Holman JA, Toth N, Shick K (1996) Paleoindian cave dwellers in the Amazon: the peopling of the Americas. Science 272(5260):373–384. https://doi.org/10.1126/science.272.5260.373

Scarano FR, Ceotto P (2015) Brazilian Atlantic forest: impact, vulnerability, and adaptation to climate change. Biodivers Conserv 24(9):2319–2331. https://doi.org/10.1007/s10531-015-0972-y

Scott JC (2017) Against the grain: a deep history of the earliest states. Yale University Press, New Haven

Smith BD (2007) Niche construction and the behavioral context of plant and animal domestication. Evol Anthropol Issues News Rev 16(5):188–199. https://doi.org/10.1002/evan.20135

Smith BD (2011) General patterns of niche construction and the management of 'wild' plant and animal resources by small-scale pre-industrial societies. Phil Trans R Soc Lond B 366(1566):836–848. https://doi.org/10.1098/rstb.2010.0253

Soares-Filho BS, Nepstad DC, Curran LM, Cerqueira GC, Garcia RA, Ramos CA, Voll E, McDonald A, Lefebvre P, Schlesinger P (2006) Modelling conservation in the Amazon basin. Nature 440:520–523. https://doi.org/10.1038/nature04389

Souza M (2009) História da Amazônia. Editora Valer, Manaus

Sutherland WJ (2003) Parallel extinction risk and global distribution of languages and species. Nature 423(6937):276–279. https://doi.org/10.1038/nature01607

Tilman D (2012) Biodiversity & environmental sustainability amid human domination of global ecosystems. Daedalus 141(3):108–120. https://doi.org/10.1162/DAED_a_00166

Tyukavina A, Hansen MC, Potapov P, Parker D, Okpa C, Stehman SV, Kommareddy I, Turubanova S (2018) Congo Basin forest loss dominated by increasing smallholder clearing. Sci Adv 4(11):eaat2993. https://doi.org/10.1126/sciadv.aat2993

Van Gemerden BS, Olff H, Parren MPE, Bongers F (2003) The pristine rain forest? Remnants of historical human impacts on current tree species composition and diversity. J Biogeogr 30(9):1381–1390. https://doi.org/10.1046/j.1365-2699.2003.00937.x

Vandermeer J, Perfecto I (2014) Paradigms lost: tropical conservation under late capitalism. In: Hecht SB, Morrison KD, Padoch C (eds) The social lives of forests: past, present, and future of woodland resurgence. The University of Chicago Press, Chicago, pp 114–128

Venter O, Sanderson EW, Magrach A, Allan JR, Beher J, Jones KR, Possingham HP, Laurance WF, Wood P, Fekete BM, Levy MA, Watson JEM (2016) Sixteen years of change in the global terrestrial human footprint and implications for biodiversity conservation. Nat Commun 7:12558. https://doi.org/10.1038/ncomms12558

Vijay V, Pimm SL, Jenkins CN, Smith SJ (2016) The impacts of oil palm on recent deforestation and biodiversity loss. PLoS One 11(7):e0159668. https://doi.org/10.1371/journal.pone.0159668

Watling J, Shock MP, Mongelo GZ, Almeida FO, Kater T, Oliveira PE, Neves EG (2018) Direct archaeological evidence for Southwestern Amazonia as an early plant domestication and food production centre. PLoS One 13(7):e0199868. https://doi.org/10.1371/journal.pone.0199868

Watson JEM, Evans T, Venter O, Williams B, Tulloch A, Stewart C, Thompson I, Ray JC, Murray K, Salazar A, McAlpine C, Potapov P, Walston J, Robinson JG, Painter M, Wilkie D, Filardi C, Laurance WF, Houghton RA, Maxwell S, Grantham H, Samper C, Wang S, Laestadius L, Runting RK et al (2018) The exceptional value of intact forest ecosystems. Nat Ecol Evol 2(4):599–610. https://doi.org/10.1038/s41559-018-0490-x

Wiersum KF (1997) From natural forest to tree crops, co-domestication of forests and tree species, an overview. Nether J Agric Sci 45(4):425–438

Wiersum KF (2004) Forest gardens as an 'intermediate' land-use system in the nature–culture continuum: characteristics and future potential. Agr Syst 61–62(1–3):123–134. https://doi.org/10.1023/B:AGFO.0000028994.54710.44

Wilson EO (2016) Half-earth: our planet's fight for life. WW Norton & Company, New York

Yasuoka H (2013) Dense wild yam patches established by hunter-gatherer camps: beyond the wild yam question, toward the historical ecology of rainforests. Hum Ecol 41(3):465–475. https://doi.org/10.1007/s10745-013-9574-z

Zarrillo S, Gaikwad N, Lanaud C, Powis T, Viot C, Lesur I, Fouet O, Argout X, Guichoux E, Salin F, Solorzano RL, Bouchez O, Vignes H, Severts P, Hurtado J, Yepez A, Grivetti L, Blake M, Valdez F (2018) The use and domestication of *Theobroma cacao* during the mid-Holocene in the upper Amazon. Nat Ecol Evol 2:1879–1888. https://doi.org/10.1038/s41559-018-0697-x

Chapter 4
Protected Areas and Food Security: Unravelling the Issues

Winy Vasquez and Terry Sunderland

4.1 Introduction

As of 2016, one in every nine people around the world is chronically undernourished, representing a worrying precedent in a historically declining statistic (FAO et al. 2017). This figure has been the focus of several international policies, targets and decrees that have brought the international community together to try and combat the global issue of alleviating hunger and malnutrition. At the same time, the International Union for Conservation of Nature (IUCN) estimates that around 15% of the world's land area and 7% of the world's oceans have been designated as protected areas (PAs), falling short of the 2020 Aichi Biodiversity targets of 17% of terrestrial PAs and 10% of oceanic PAs (IUCN 2019). PAs and the global hunger statistics may, at first glance, seem like two unrelated global challenges, or even competing interests, but they are in fact intrinsically interlinked.

Combating malnutrition is a critical development objective due to the long-term and far-reaching health and socioeconomic implications of malnutrition such as, compromised cognitive development in children (Cawthorn and Hoffman 2015), childhood stunting (Fa et al. 2015; Nielsen et al. 2018) and increased susceptibility to non-communicable diseases (Popkin 2001; Vinceti et al. 2013; Savage et al. 2019). Biodiversity conservation is likewise an important objective, due to the rapid and ongoing depletion of species and concomitant habitat destruction occurring worldwide (Morales-Hidalgo et al. 2015). Similar to malnutrition, biodiversity loss has far-reaching impacts, which negatively impacts both humans and nature. While

W. Vasquez
Faculty of Forestry, University of British Columbia, Vancouver, BC, Canada

T. Sunderland (✉)
Faculty of Forestry, University of British Columbia, Vancouver, BC, Canada

Centre for International Forestry Research, Bogor, Indonesia
e-mail: terry.sunderland@ubc.ca

C. Baldauf (ed.), *Participatory Biodiversity Conservation*,
https://doi.org/10.1007/978-3-030-41686-7_4

interest in PAs as an effective means to safeguard biodiversity continues to grow, so has the parallel global movement to eradicate hunger and malnutrition.

In 1974, at the inaugural World Food Conference, 135 participating countries issued the Universal Declaration on the Eradication of Hunger and malnutrition which declared that '*[e]very man, woman and child has the inalienable right to be free from hunger and malnutrition in order to develop fully and maintain their physical and mental faculties*' (UN General Assembly, 1975, art. 1). Yet this 'inalienable right' to be free from hunger is still, to this day, denied for many rural populations that live within or adjacent to PAs, where there is a strong emphasis on enforcement and restricted access. In 1996, the global community once again came together to reaffirm the rights to food and freedom from hunger during the World Food Summit which resulted in highlighting food security as the new global goal. During this summit, food security was defined as '*exists[ing] when all people, at all times, have physical and economic access to sufficient, safe and nutritious food that meets their dietary needs and food preferences for an active and healthy life*' (World Food Summit, 1996). Food insecurity has plagued many countries around the world and several strategies have been employed over the years to help fight it but the ability of natural resources to help combat food insecurity remains understudied and underutilized.

PAs can lead to food insecurity through a variety of pathways such as a loss of direct access to the harvesting of wild foods (Nakamura and Hanazaki 2016), loss of livestock due to predation by wildlife (Banerjee 2012; Givá and Raitio 2017), loss of access to bodies of water used for irrigation or drinking water (Adhikari et al. 2009; N'Danikou et al. 2017), loss of fuelwood for cooking (Banerjee 2012), loss of traditional knowledge (Turner and Turner 2008; Desmet 2016), and loss of access to markets and increased food prices due to tourism (Rosendo et al. 2011; Bennett and Dearden 2014), to name a few. This chapter will therefore look at how a move away from the traditional fortress conservation approach can not only help increase food security but also lead to more effective conservation outcomes.

Box 4.1 Selected Policies and Legislative Framework Related to Food Security

1975: The IUCN passed the Kinshasa Resolution on the protection of the 'traditional ways of life' and called on governments to halt the displacement and relocation of people due to PA (Adams and Hutton 2007)

2003: Durban Action Plan, outcome 5 'The rights of indigenous peoples, including mobile indigenous peoples, and local communities are secured in relation to natural resources and biodiversity conservation'

2004: Convention on biological diversity called for the recognition of 'the economic and socio-cultural costs and impacts arising from the establishment and maintenance of protected areas, particularly for indigenous and local communities, and (an adjustment of) policies to ensure that such costs and impacts—including the cost of livelihood opportunities forgone—are equitably compensated'

2007: Establishment of the United Nations Declaration on the Rights of Indigenous Peoples (UNDRIP) affirmed the rights, survival, dignity and well-being of Indigenous people as well as safeguard the individual and collective rights of Indigenous people that may not be addressed by other human rights charters.

2010: Aichi Biodiversity Target 11 aims to protect 17% of terrestrial and inland water, and 10% of coastal and marine areas to be conserved by 2020 through '*effectively and equitable managed, ecologically representative and well-connected systems of protected areas*'

2012: UN Zero Hunger Challenge, which calls for sustainable food systems, an end to rural poverty, adaptation of all food systems to eliminate loss or waste of food, increase access to adequate food and healthy diets for all people all year round and finally, for an end to malnutrition in all its forms.

4.2 A Brief History of Conservation

The first official park, Yellowstone National Park, was established in the United States in 1872, and its creation set the stage for other countries to soon follow suit (Adams and Hutton 2007). Yellowstone National Park thus became the model park that other countries looked to when creating their own PAs, and so began the century long legacy of separating 'wild land' and 'wilderness' from human existence and human use (Shafer 2015). Yellowstone National Park is a prime example of the development of a 'Western' dichotomous view of wilderness as being separate from humans, since the park was established under the pretext of setting aside 'wilderness' for the enjoyment and benefit of the public, yet it deprived Native Americans from this supposed benefit by barring them from living and accessing their traditional territory (Shafer 2015). As the PA movement around the world was gaining ground, so was the Western perspective of nature as 'untouched', 'uninhabited' and 'unaltered' and this led to conflict and human rights abuses around the world as local inhabitants were excluded in the name of game conservation, and later under the guise of broader biodiversity conservation (Shafer 2015). In Canada, the establishment of the country's first National Pak, Banff, mirrored the exclusionary behaviour that followed the establishment of Yellowstone National Park and other PAs in the United States. Banff National Park was also an example of the 'Western' understanding of wilderness as being unaltered by humans, when 'pristine' wilderness as a concept is being contested (Binnema and Niemi 2006; West et al. 2006; Shafer 2015). The Western idea of 'pristine nature' was only possible because it failed to recognize how indigenous groups have been altering the landscape for centuries and classified these altered landscapes as 'natural' and 'wild' (Adams and Hutton 2007; Shafer 2015; Massé 2016; Anaya and Espírito-Santo 2018).

PAs have been negatively impacting the food security of local communities since their inception, as can be seen in the case of Banff National Park. There, sportsmen continuously campaigned to restrict the hunting of the local indigenous groups because, in their eyes, subsistence hunting for food did not align with the hunting for sport and trophy hunting on which they so heavily prided themselves (Binnema and Niemi 2006). In common with Banff National Park, many PAs around the world have had a long history of decoupling food security from the natural world by failing to understand the important role that natural resources play in the healthy and nutritious diets of rural populations (Powell et al. 2015). This sentiment has also been expressed by local populations living in or around PAs and has caused tensions and violence in many areas (West et al. 2006). In Cantanhez Forest National Park, in Guinea-Bissau, West Africa, for example, interviews with women living in the park revealed that they felt the park was directly responsible for malnutrition in their communities due to a lack of compensation when devastating crop raiding occurred in their fields (Costa et al. 2017).

The establishment of PAs has often been characterized by the forced relocation of those who lived within the newly dotted park boundaries and the beginning of extraction and land-use regulations as PAs were, at first, primarily established for the conservation of game, not biodiversity (Binnema and Niemi 2006). This model of PAs establishment alongside the, often forced, relocation of indigenous populations proved to be a standard approach for decades to come and has led to many negative impacts on the livelihoods and well-being of hundreds of communities. The number of documented abuses of power and human rights due to the establishment, management and policing of PAs has been so prolific that the seriousness of the problem was recognized at an international level as early as 1982 at the Third World Park Congress (WPC), but a new agreement highlighting the problem wasn't reached until the 5th WPC in 2003 (IUCN 2005; Adams and Hutton 2007). At this event, the Durban Accord was established to represent a shift in thinking that recognized the need to involve indigenous communities and address their needs in the context of PAs (Adams and Hutton 2007). Yet, despite mounting evidence on human rights violations with regard to food access in PAs, land annexation for PAs has continued to grow while the enforcement of the right to food lags behind. While the conservation community has made increasing strides in recent decades to move away from the 'fortress conservation' approaches and towards community-based natural resource managements, that take into account local concerns and livelihoods, the integration of livelihoods and food security into biodiversity protection still has a long way to go.

PAs today are often established for biological conservation but this wasn't always the case. As already mentioned, many of the PAs that were set aside in the nineteenth century were predominantly undertaken in order to protect game for hunting and state access to valuable natural resources, thus consolidating the power over land rights and natural resources into the hands of a few wealthy stakeholders (West et al. 2006; Lunstrum and Ybarra 2017). In the following decades, however, the rationale behind the establishment of PAs has shifted considerably as ideas of 'wilderness' have changed alongside the development of fields such as conservation biology, which were fuelled by concerns over increasing global biodiversity loss. The IUCN itself has undergone several changes in how it approaches PAs. In 1994, the IUCN released six PA categories that classified them on the basis of management objectives, which some have critiqued as being incongruent with the ultimate goal of biodiversity conservation due to their narrow focus on management (Boitani et al. 2008). Fast forward more than two decades later and the IUCN PA categories have expanded their definitions to go beyond management objectives and are increasingly recognizing the need to incorporate rights-based approaches, social inclusion, indigenous rights, livelihoods and benefit-sharing into their mandate (IUCN 2019).

The right to food and the importance of access to ancestral land have been, in recent years, gaining ground in many countries as cases have been brought before the courts contesting land and livelihood rights. One such example occurred in Botswana and resulted in a ruling against the government and their eviction of Bushmen from the Central Kalahari Game Reserve, proving that the displacement of the local Bushmen was 'unlawful and unconstitutional' and acknowledged their right to live on their ancestral land (Adams and Hutton 2007). In Canada, many indigenous groups have also brought cases against the government regarding land rights that have, among other things, impacted their access to traditional food systems by restricting traditional hunting and fishing practices (Desmarais and Wittman 2014).

4.3 How Do Forests Play an Important Role in Diets and Nutrition?

PAs are often presented as a haven for biodiversity conservation, yet the link between biodiversity and nutrition is not often recognized. The body of literature linking biodiversity and natural resources to healthy diets has been growing in recent years, renewing interest in how PAs impact the 'right to food'. A study of the diets of children in 21 African countries, for example, showed a positive relationship between tree cover and dietary diversity among the diet of children, thus highlighting the importance of tree cover for more diverse and nutritious diets (Ickowitz et al. 2014). Food insecurity has for years been fought with increased production and a race to obtain higher yields in agricultural crops, yet an increase in food supply has not resulted in an increase in food security or nutritious diets (Fischer et al. 2017; Ickowitz et al. 2019). It is therefore important to begin to look to other strategies to help mitigate food insecurity and one of these strategies lies in access to natural resources.

4.3.1 Ecosystem Services (Related to Food Security): Water Regulation, Pollination, Soil Erosion Control, Nutrient Cycling and So On

Ecosystem Services are the benefits that humans receive from the ecosystem, ranging from water regulation and protection of soil erosion to the provisioning of wild foods and pollination (Richardson 2010; Sunderland 2011; IUCN 2013; Mcneely 2016; Reed et al. 2017). Ecosystems services can be divided into four major categories: provisioning, regulating, supporting and cultural; all of which play a role in food security. Provisioning ecosystem services can come in the form of direct provision of foods, medicinal plants and fuelwood, discussed in later sections (Fernandez-Llamazares et al. 2017; Ahammad et al. 2019). Regulating ecosystem services on the other hand are indirect benefits such as water purification, pest and disease control and pollination (Fernandez-Llamazares et al. 2017; Ahammad et al. 2019). Cultural ecosystem services such as aesthetics and spiritual practices can also play important roles in food security as they can inform behaviours and form the basis for how ecosystems are valued and thus protected (Richardson 2010; Ahammad et al. 2019). Lastly, ecosystems services can play a supporting roles by controlling soil erosion, aiding in nutrient cycling and soil formation, all of which are important services for food production (Richardson 2010). A healthy ecosystem, like those envisioned within PAs, can therefore play a key role in the pursuit of food security for communities living both inside and in proximity to PAs. In marine ecosystems, for example, a loss of biodiversity can lead to a loss of ecosystem services like fisheries, nursery habitat, filtration and detoxification services (Worm et al. 2006).

4.3.2 Contribution of Wild Foods to Diets

Another important contribution of forests to food security is in the form of the direct provisioning of wild foods such as edible plants, nuts, seeds and wild meat, hereinafter referred to as bushmeat. Research has demonstrated that many rural populations that live in or around forested areas rely, to varying degrees, on the harvesting of wild foods to help meet their dietary needs (Boedecker et al. 2014; Rowland et al. 2017; Sunderland et al. 2013; Sunderland 2011). A 2015 global comparative analysis across three continents found that 77% of the households surveyed engaged in wild food collection, showing that wild food harvesting is an integral part of many households in developing countries (Hickey et al. 2016). The harvesting of wild foods can contribute to food security by allowing rural dwellers to access nutritious foods when they may otherwise not have other sources of food (Boedecker et al. 2014). Access to wild foods is also an important part of food security as it can help mitigate hardships brought on by internal and external shocks such as droughts, war, illness and/or failing crops (Pouliot and Treue 2013; Clements et al. 2014). Access to wild foods can also bring resilience to traditional agricultural systems by providing a safety net in case of crop failures, pests infestations or crop raiding by animals, a common occurrence in and around PAs (Nyahongo et al. 2009; Pouliot and Treue 2013; Schulte-Herbrüggen et al. 2013; Wunder et al. 2014; Cawthorn and Hoffman 2015; Galway et al. 2018).

While an agricultural system can provide a family with a few staple food crops and help fulfil the daily caloric requirements of an individual, it often doesn't adequately provide a diverse and nutritious diet when compared to that possible when supplemented with locally available wild foods (Nakamura and Hanazaki 2016; Fischer et al. 2017). Studies have shown that increased agricultural production has, in some cases, actually led to lower quality diets that are composed of calorie-rich food which lack important micronutrients such as iron, zinc and vitamin B12 (Cawthorn and Hoffman 2015; Powell et al. 2015; Sunderland et al. 2013). Wild foods don't always contribute a large percentage of a rural household's diets, but they have been found, in several studies, to contribute essential vitamins and minerals (Powell et al. 2015; Asprilla-Perea and Díaz-Puente 2019). While studies on the nutritional content of wild varieties remain limited, there is evidence to suggest that wild varieties can carry higher nutritional content when compared to domestic varieties (Burlingame et al. 2009; Heywood 2011; Savage et al. 2019). Furthermore, the higher nutritional value and resiliency of native wild foods will also prove to be of great importance in the face of climate change (Savage et al. 2019).

Harvesting wild foods can increase dietary diversity and help combat micronutrient deficiencies also known as 'hidden hunger' (Ickowitz et al. 2014; Fa et al. 2015; Nielsen et al. 2018). Micronutrient deficiency is an important aspect of malnutrition that can have dire consequences in vulnerable sectors of the population such as young children and can lead to childhood stunting, which has life-long consequences (Temsah et al. 2018). A recent study looking at 15 sub-Saharan African countries found a negative relationship between deforestation and dietary diversity (Galway et al. 2018). Blaney et al. (2009) assessing the contribution of natural resources to the nutritional status of the local population in a protected area in Gabon found that the consumption of natural resources by children aged 5–9 was the best predictor for nutritional status. While foods from natural resources only contributed to 12% of the energy requirements of villagers of the Gamba Complex of Gabon, they contributed an estimated 82% of protein, 36% of vitamin A and 20% of iron requirements and were found to be more nutrient dense than non-wild foods (Blaney et al. 2009). In Canada, decreased access, due to wildlife declines or harvesting restrictions, can lead to micronutrient deficiencies in Inuit communities as Caribou constitutes the primary source of many micronutrients such as zinc, copper and B12 (Kenny et al. 2018).

Hunting for bushmeat has long been a controversial subject due to concerns over unsustainable hunting practices and wildlife depletion but bushmeat hunting is also important in helping rural households to achieve food security (Fa et al. 2009; Nyahongo et al. 2009; Rentsch and Damon 2013; Golden et al. 2014; Cawthorn and Hoffman 2015; Reuter et al. 2016; Nielsen et al. 2018). In the Abun region of West Papua, Indonesia, hunting has proved to be an important factor in fighting food insecurity, as bushmeat accounted for 49% of the diets of respondents

(Pattiselanno and Lubis 2014). Bushmeat hunting around the world remains an important source of protein for many rural households and can provide vulnerable populations such as children with important micronutrients (Golden et al. 2011; Van Vliet et al. 2015). In Madagascar, a study on biodiversity and child nutrition found that a reduction in bushmeat consumption, either by restricted access or by wildlife depletion, could lead to a predicted 29% increase in children with anaemia and a tripling of anaemia in children in the poorest households (Golden et al. 2011). One analysis of 24 countries across Latin America, Asia and Africa found that, on average, 89% of the income obtained from bushmeat was for the households' own consumption, dismissing a common falsehood that wildlife hunting is undertaken primarily for economic benefit (Nielsen et al. 2018). The same study also found that households in Latin America were the most reliant on bushmeat hunting, particularly as an opportunistic activity. Gardner and Davies (2014) echoed this finding and also found bushmeat hunting in Madagascar to be an opportunistic activity carried out when users visit forests to extract other resources. That being said, bushmeat hunting in order to supply urban markets, where bushmeat is seen as a delicacy or a luxury, has been increasing as societies become urbanized (Cawthorn and Hoffman 2015).

One strategy to help relieve some of the hunting pressure on wildlife could be to switch households from wild to domesticated meat sources but this is not always feasible due to strong preference or cultural believes attached to bushmeat consumption in some areas, as well as a preference to use domesticated animals for transport and labour rather than as a source of food (Gardner and Davies 2014). Switching households to domesticated meat could also present a conservation challenge as increased demands for domesticated meat could mean increased demand for pasture lands, which often comes at the expense of forests (Nasi et al. 2011). Understanding the dynamics of why, when and where bushmeat hunting occurs can help to devise coherent policies and practices that can help to reduce commercial bushmeat hunting while still maintaining the flexibility needed to support the food security of rural households (Rentsch and Damon 2013; Van Vliet et al. 2015)

Bushmeat can be such an important aspect of people's diet that even when considerable risk of illness from a zoonic disease exists, people are still willing to risk eating bushmeat. This was the case in Chamba, Zambia, where an anthrax outbreak in humans was reported near an area where 85 hippopotamuses died of suspected anthrax, and yet 23% of those interviewed stated they would continue to eat meat from dead hippopotamus they came across, due to food shortage, lack of meat, hunger and protein shortage (Lehman et al. 2017). This anthrax outbreak shows how chronic food insecurity can lead to decisions that can severely impact the health and well-being of people.

4.3.3 Bioenergy

An often overlooked, but crucial aspect of a healthy diet is access to a source of energy in order to properly cook foods and access important minerals and vitamins that may otherwise be inaccessible. One of the most common sources of energy in rural areas is fuelwood, and some estimates point to approximately 2.4 billion people around the world relying on fuelwood as their main source of cooking fuel (Galway et al. 2018). Fuelwood and charcoal production are both important for cooking and an important source of income for many rural households (McElwee 2010; Pouliot and Treue 2013; Kelboro and Stellmacher 2015). One study on the impacts of PAs on local livelihoods in Cam Xuyen, Vietnam, found that nearly 100% of households in the study area relied on fuelwood as their main energy source (McElwee 2010). Another study in Jharkhand, India, relayed similar findings by demonstrating that fuelwood was 'by far the most important forest product' in the study site (Belcher et al. 2015).

4.3.4 Health: Food as Medicine

Medicinal plants have been playing an important role in human health and nutrition since time immemorial and despite a recent 'Western' division between food, medicine and health, they continue to be an important contributor to health and well-being for many communities (Heywood 2011). Access to wild foods is therefore important for human health, since nutrition and health are intrinsically linked. The impact of a loss of medicinal plants and nutritious diets can be seen in many Indigenous communities that have undergone nutritional transitions. For example, Indigenous communities in Canada (Binnema and Niemi 2006; Damman et al. 2008), Argentina (Damman et al. 2008), Sri Lanka (Weerasekara et al. 2018), the Eastern Mediterranean (Heywood 2011) and Borneo (Dounias et al. 2007) have all undergone nutritional transitions away from their traditional diets. This dietary shift towards a narrower range of foods that are higher in fat, sugar, salt and refined carbohydrates has led to a documented increase in the prevalence of non-communicable diseases like cardiovascular disease and diabetes in the affected populations (Popkin 2001; Albala et al. 2002; Kuhnlein et al. 2004; Damman et al. 2008; Lourenço et al. 2008; Savage et al. 2019). This has been due to both an increase in a nutritionally poor diet that makes individuals more susceptible to disease and illness and a decrease in access to traditional medicinal plants. One study looking at indigenous forest foods and HIV/AIDS in West and Central Africa discusses how the nutritional composition of forest foods could help to diversify cereal and tuber-based diets, increase micronutrients and help maintain an optimal production of antibodies—crucial in HIV/AIDS vulnerable communities (Kengni et al. 2004).

4.4 Reconciling Rights and Access to Food and Dietary Diversity

The manner in which protected areas, whether marine or terrestrial, are established and managed can lead to a variety of impacts on the local community, and it is therefore of paramount importance to carefully consider these impacts. In the Philippines, for example, marine protected areas (MPAs) were found to be positively associated with the dietary diversity of children who lived within 2 km of the MPA (Alva et al. 2016). In this case, MPAs had a net positive impact on the food security of coastal communities through the protection of fish biodiversity and highlights an instance in which both biodiversity and food security were simultaneously achieved. In Brazil, however, the establishment of PAs in the Minas Gerais state led to negative effects on the food security and livelihoods of the local communities (Anaya and Espírito-Santo 2018). When the creation of PAs jeopardizes the livelihoods of people, conflict often arises which can actually lead to 'double unsustainability' in which biodiversity conservation and food security are both compromised (Anaya and Espírito-Santo 2018). The creation or enforcement of existing PAs should therefore look to achieve win-win scenarios in which both conservation and livelihood goals are achieved. A meta-analysis of 55 PAs in developing countries found the variable that most influenced the level of compliance with PA policies was the level of involvement of local communities in the decision-making process (Andrade and Rhodes 2012). This meta-analysis therefore gives further credence to the call for rights-based approaches which recognizes and respects the rights of local communities. Chhatre and Agrawal (2009) likewise found that higher levels of involvement and decision-making power of local communities led to favourable conservation outcomes. Therein lies an opportunity to rethink how PAs are enacted and managed in order to support both biodiversity conservation and food security.

As stated previously, many PAs around the world have resulted in the loss of land rights and food access for local populations which has in turn negatively impacted the diets and nutrition of nearby communities. In order to revert some of these impacts, it is important to understand how management strategies can lead to food insecurity. Enlisting new strategies to alleviate food insecurity and biodiversity loss will thus require the involvement of multiple disciplines to contribute innovative ways forward (Brockington et al. 2006; Timko and Satterfield 2008). A move towards an increased recognition of synergies, rather than trade-offs, between food security and biodiversity conservation presents an opportunity for the emergence of new conservation frameworks that build on rights-based approaches, food sovereignty principles, and participatory conservation to rethink how PA enactment, management and policing is approached.

Rights-based approaches to conservation will be one key instrument in moving towards more salient conservation policies that integrate the fundamental 'right to food' by helping to identify rights-holders and duty-bearers to better inform PA management (Young et al. 2004; He and Cliquet 2014). Adopting a rights-based approach to conservation will present its own set of challenges, such as funding, lack of expertise and/or government capacity and competing rights, but it is a necessary step forward that can help to increase both conservation and food security (He and Cliquet 2014; Kraak 2018). In some cases, a rights-based approach will require the dissemination of power within PAs in favour of more egalitarian, bottom-up approaches such as community-based conservation projects and livelihood-based conservation (Campese et al. 2009), in order to achieve the dual goal of conservation and food security. A meta-analysis of 165 PAs found that PAs that were associated with a positive socioeconomic outcome were more likely to also report a positive conservation outcome and thus demonstrated that conservation and food security goals are not antagonistic (Oldekop et al. 2016). Rights-based approaches can also increase the resiliency of both humans and nature by supporting both social and environmental justice through collaboration and shared responsibility (Walsh-Dilley et al. 2016). Using the rights-based approach to empower local communities to make their own management decisions around harvesting, logging and other resources practices can actually increase conservation outcomes as an increase in rights and responsibilities decreases unsustainable harvesting practices (Nielsen et al. 2018). In this way, locals can act as 'gate keepers' by deterring outsiders who want to undertake unsustainable harvesting practices inside of PAs (Adhikari et al. 2009; Kubo and Supriyanto 2010; Nielsen et al. 2018)

Looking at PA management through a food sovereignty lens will also help to promote inclusive and socially responsible management strategies while building the resiliency of communities by decreasing dependency on outside sources such as governments and NGOs (Zavaleta et al. 2017). Food sovereignty will be of particular importance for Indigenous communities, who have deep historical ties to their lands and resources (Turner et al. 2011; Kuhnlein et al. 2013) and whose traditional territories are estimated to overlap with at least 40% of PAs (Garnett et al. 2018). Recognition of food sovereignty could therefore help to better inform PA management with respect to traditional food systems and harvesting practices (Sylvester et al. 2016). Food sovereignty is helping to reshape the political arena under which decisions and values are changing when it comes to how food is produced, accessed and consumed (Desmarais and Wittman 2014), and while food sovereignty is in its infancy stage in many countries, this could help lay the groundwork for a paradigm shift in how conservation is approached.

Despite the fact that PAs have often played an antagonistic role in the lives of local communities, recent shifts to more community-based conservation practices signals a paradigm shift that could see more inclusive management practices. In the Gunung Halimun-Salak National Park in Indonesia, an effort to move away from the 'fences and fines' approach and towards a 'participatory' approach allowed room for relationship building between frontline park staff and villagers which resulted in the villagers refraining from illegal logging and mining practices (Kubo and Supriyanto 2010). This behavioural shift by local villagers emphasizes the importance of trust and relationship building, as frontline park staff attributed this shift in the local villagers to increased trust rather than to economic incentives (Kubo and Supriyanto 2010).

4.5 Concluding Remarks

As the contribution of forests and tree-based systems continues to be recognized, so does the opportunity to reconcile conservation in PAs with the rights to food in these spaces. With the increasingly growing demand to conserve more land and seascapes and reach the goals set out by global treaties, it is now more important than ever to move forward with more inclusive management programmes that don't jeopardize human livelihoods. As the amount of land that is set aside for the creation or expansion of PAs continues to grow, so does the opportunity to recognize and rework broken management schemes that do not accurately reflect the social cost of conservation, the burden of which is most heavily felt by the poor and disfranchised parts of the population. While the recognition of rights-based approaches to conservation and rights to food will help to increase food security, it is not the only solution and will need to work in cohort with other strategies to help alleviate food insecurity around the world.

References

Adams WM, Hutton J (2007) People, parks and poverty: political ecology and biodiversity conservation. Conserv Soc 5:147–183

Adhikari KR, Tan Y-C, Lai J-S, Pant D (2009) Irrigation intervention: a strategy for conserving biodiversity and improving food security in Royal Chitwan National Park buffer zone, Nepal. Irrig Drain 58:522–537

Ahammad R, Stacey N, Sunderland TCH (2019) Use and perceived importance of forest ecosystem services in rural livelihoods of Chittagong Hill Tracts, Bangladesh. Ecosyst Serv 35:87–98. https://doi.org/10.1016/j.ecoser.2018.11.009

Albala C, Vio F, Kain J, Uauy R (2002) Nutrition transition in Chile: determinants and consequences. Public Health Nutr 5:123–128. https://doi.org/10.1079/phn2001283

Alva S, Johnson K, Jacob A et al (2016) Marine protected areas and children's dietary diversity in the Philippines. Popul Environ 37:341–361. https://doi.org/10.1007/s11111-015-0240-9

Anaya FC, Espírito-Santo MM (2018) Protected areas and territorial exclusion of traditional communities: Analyzing the social impacts of environmental compensation strategies in Brazil, Ecol Soc 23. https://doi.org/10.5751/ES-09850-230108

Andrade GSM, Rhodes JR (2012) Protected areas and local communities: an inevitable partnership toward successful conservation strategies? Ecol Soc 17. https://doi.org/10.5751/ES-05216-170414

Asprilla-Perea J, Díaz-Puente JM (2019) Importance of wild foods to household food security in tropical forest areas. Food Secur 11:15–22

Banerjee A (2012) Is wildlife tourism benefiting Indian protected areas? A survey. Curr Issues Tour 15:211–227. https://doi.org/10.1080/13683500.2011.599367

Belcher B, Achdiawan R, Dewi S (2015) Forest-based livelihoods strategies conditioned by market remoteness and forest proximity in Jharkhand, India. World Dev 66:269–279. https://doi.org/10.1016/j.worlddev.2014.08.023

Bennett NJ, Dearden P (2014) Why local people do not support conservation: community perceptions of marine protected area livelihood impacts, governance and management in Thailand. Mar Policy 44:107–116. https://doi.org/10.1016/j.marpol.2013.08.017

Binnema T, Niemi M (2006) "Let the Line Be Drawn Now": wilderness, conservation, and the exclusion of aboriginal people from Banff National Park in Canada. Environ Hist Durh N C 11:724–750

Blaney S, Beaudry M, Latham M (2009) Contribution of natural resources to nutritional status in a protected area of Gabon. Food Nutr Bull 30:49–62. https://doi.org/10.1177/156482650903000105

Boedecker J, Termote C, Assogbadjo AE et al (2014) Dietary contribution of wild edible plants to women's diets in the buffer zone around the Lama forest, Benin – an underutilized potential. Food Secur 6:833–849. https://doi.org/10.1007/s12571-014-0396-7

Boitani L, Cowling RM, Dublin HT et al (2008) Change the IUCN protected area categories to reflect biodiversity outcomes. PLoS Biol 6:0436–0438. https://doi.org/10.1371/journal.pbio.0060066
Brockington D, Igoe J, Schmidt-Soltau K (2006) Conservation, human rights, and poverty reduction. Conserv Biol 20:250–252. https://doi.org/10.1111/j.1523-1739.2006.00335.x
Burlingame B, Charrondiere R, Mouille B (2009) Food composition is fundamental to the cross-cutting initiative on biodiversity for food and nutrition. J Food Compos Anal 22:361–365
Campese J, Sunderland T, Greiber T, Oviedo G (eds) (2009) Rights-based approaches: exploring issues and opportunities for conservation. CIFOR and IUCN, Bogor
Cawthorn DM, Hoffman LC (2015) The bushmeat and food security nexus: a global account of the contributions, conundrums and ethical collisions. Food Res Int 76:906–925. https://doi.org/10.1016/j.foodres.2015.03.025
Chhatre A, Agrawal A (2009) Trade-offs and synergies between carbon storage and livelihood benefits from forest commons. Proc Natl Acad Sci U S A 106:17667–17670. https://doi.org/10.1073/pnas.0905308106
Clements T, Suon S, Wilkie DS, Milner-Gulland EJ (2014) Impacts of protected areas on local livelihoods in Cambodia. World Dev 64:S125–S134. https://doi.org/10.1016/j.worlddev.2014.03.008
Costa S, Casanova C, Lee P (2017) What does conservation mean for women? The case of the Cantanhez Forest National Parl. Conserv Soc 15:168–178
Damman S, Eide WB, Kuhnlein HV (2008) Indigenous peoples' nutrition transition in a right to food perspective. Food Policy 33:135–155. https://doi.org/10.1016/j.foodpol.2007.08.002
Desmarais AA, Wittman H (2014) Farmers, foodies and first nations: getting to food sovereignty in Canada. J Peasant Stud 41:1153–1173. https://doi.org/10.1080/03066150.2013.876623
Desmet E (2016) Identifying rights-holders in natural resource regimes: a critical assessment of the Peruvian protected areas legislation. Hague J Rule Law 8:135–154. https://doi.org/10.1007/s40803-016-0027-9
Dounias E, Selzner A, Koizumi M, Levang P (2007) From sago to rice, from forest to town: the consequences of sedentarization for the nutritional ecology of Punan former hunter-gatherers of Borneo. Food Nutr Bull 28:294–302. https://doi.org/10.1177/15648265070282S208
Fa JE, Albrechtsen L, Johnson PJ, Macdonald DW (2009) Linkages between household wealth, bushmeat and other animal protein consumption are not invariant: evidence from Rio Muni, Equatorial Guinea. Anim Conserv 12:599–610. https://doi.org/10.1111/j.1469-1795.2009.00289.x
Fa JE, Olivero J, Real R et al (2015) Disentangling the relative effects of bushmeat availability on human nutrition in Central Africa. Sci Rep 5:1–8. https://doi.org/10.1038/srep08168
FAO, IFAD, UNICEF et al (2017) The state of food security and nutrition in the world 2017. Building resilience for peace and food security
Fernandez-Llamazares A, Cabeza M, Terraube J (2017) The role of protected areas in supporting human health: a call to broaden the assessment of conservation outcomes. Sci Direct 25:50–58. https://doi.org/10.1016/j.cosust.2017.08.005
Fischer J, Abson DJ, Bergsten A et al (2017) Reframing the food–biodiversity challenge. Trends Ecol Evol 32:335–345. https://doi.org/10.1016/j.tree.2017.02.009
Galway LP, Acharya Y, Jones AD (2018) Deforestation and child diet diversity: a geospatial analysis of 15 sub-Saharan African countries. Heal Place 51:78–88. https://doi.org/10.1016/j.healthplace.2018.03.002
Gardner CJ, Davies ZG (2014) Rural bushmeat consumption within multiple-use protected areas: qualitative evidence from Southwest Madagascar. Hum Ecol 42:21–34
Garnett ST, Burgess ND, Fa JE et al (2018) A spatial overview of the global importance of indigenous lands for conservation. Nat Sustain 1:369–374
Givá N, Raitio K (2017) 'Parks with people' in Mozambique: community dynamic responses to human-elephant conflict at Limpopo national park. J South Afr Stud 43:1199–1214. https://doi.org/10.1080/03057070.2017.1374810
Golden CD, Fernald LCH, Brashares JS et al (2011) Benefits of wildlife consumption to child nutrition in a biodiversity hotspot. Proc Natl Acad Sci U S A 108:19653–19656. https://doi.org/10.1073/pnas.people

Golden CD, Bonds MH, Brashares JS et al (2014) Economic valuation of subsistence harvest of wildlife in Madagascar. Conserv Biol 28:234–243. https://doi.org/10.1111/cobi.12174
He M, Cliquet A (2014) Sustainable development through a rights-based approach to conserve protected areas in China. China-EU Law J 3:143–163. https://doi.org/10.1007/s12689-013-0031-7
Heywood VH (2011) Ethnopharmacology, food production, nutrition and biodiversity conservation: towards a sustainable future for indigenous peoples. J Ethnopharmacol 137:1–15
Hickey GM, Pouliot M, Smith-hall C et al (2016) Quantifying the economic contribution of wild food harvests to rural livelihoods: a global-comparative analysis. Food Policy 62:122–132. https://doi.org/10.1016/j.foodpol.2016.06.001
Ickowitz A, Powell B, Salim MA, Sunderland TCH (2014) Dietary quality and tree cover in Africa. Glob Environ Chang 24:287–294. https://doi.org/10.1016/j.gloenvcha.2013.12.001
Ickowitz A, Powell B, Rowland D et al (2019) Agricultural intensification, dietary diversity, and markets in the global food security narrative. Glob Food Sec 20:9–16. https://doi.org/10.1016/j.gfs.2018.11.002
IUCN (2005) The Durban accord, pp 219–223
IUCN (2013) Food security policies: making the ecosystem connections. Gland, Switzerland
IUCN (2019) IUCN annual report 2018. Gland, Switzerland
Kelboro G, Stellmacher T (2015) Protected areas as contested spaces: Nech Sar National Park, Ethiopia, between "local people", the state, and NGO engagement. Environ Dev 16:63–75. https://doi.org/10.1016/j.envdev.2015.06.005
Kengni E, Mbofung CMF, Tchouanguep MF, Tchoundjeu Z (2004) The nutritional role of indigenous foods in mitigating the HIV/AIDS crisis in West and Central Africa. Int For Rev 6:149–160. https://doi.org/10.1505/ifor.6.2.149.38392
Kenny T-A, Fillion M, Simpkin S et al (2018) Caribou (Rangifer tarandus) and Inuit nutrition security in Canada. EcoHealth. https://doi.org/10.1007/s10393-018-1348-z
Kraak AL (2018) Human rights-based approaches to world heritage conservation in Bagan, Myanmar: conceptual, political, and practical considerations. Int J Cult Prop 25:111–133. https://doi.org/10.1017/S0940739118000012
Kubo H, Supriyanto B (2010) From fence-and-fine to participatory conservation: mechanisms of transformation in conservation governance at the Gunung Halimun-Salak National Park, Indonesia. Biodivers Conserv 19:1785–1803. https://doi.org/10.1007/s10531-010-9803-3
Kuhnlein HV, Receveur O, Soueida R, Egeland GM (2004) Arctic indigenous peoples experience the nutrition transition with changing dietary patterns and obesity. J Nutr 134:1447–1453. https://doi.org/10.1093/jn/134.6.1447
Kuhnlein H, Erasmus B, Spigelski D, Burlingame B (eds) (2013) Indigenous peoples' food systems & well-being. FAO, Rome
Lehman MW, Craig AS, Malama C et al (2017) Role of food insecurity in outbreak of anthrax infections among humans and hippopotamuses living in a game reserve area, rural Zambia. Emerg Infect Dis 23:1471–1477. https://doi.org/10.3201/eid2309.161597
Lourenço AEP, Santos RV, Orellana JDY, Coimbra CEA (2008) Nutrition transition in Amazonia: obesity and socioeconomic change in the Suruí Indians from Brazil. Am J Hum Biol 20:564–571. https://doi.org/10.1002/ajhb.20781
Lunstrum E, Ybarra M (2017) Deploying difference: security threat narratives and state displacement from protected areas. Conserv Soc 16:114–124
Massé F (2016) The political ecology of human-wildlife conflict: producing wilderness, insecurity, and displacement in the Limpopo National Park. Conserv Soc 14:100–111. https://doi.org/10.4103/0972-4923.186331
McElwee PD (2010) Resource use among rural agricultural households near protected areas in Vietnam: the social costs of conservation and implications for enforcement. Environ Manag 45:113–131. https://doi.org/10.1007/s00267-009-9394-5
Mcneely JA (2016) Protected areas, ecosystem services, and food production. In: Gordon IJ, Prins HHT, Squire GR (eds) Food production and nature conservation: conflicts and solutions. Routledge, London; New York, pp 110–132

Morales-Hidalgo D, Oswalt SN, Somanathan E (2015) Status and trends in global primary forest, protected areas, and areas designated for conservation of biodiversity from the Global Forest Resources Assessment 2015. For Ecol Manag 352:68–77. https://doi.org/10.1016/j.foreco.2015.06.011

N'Danikou S, Vodouhe RS, Bellon MR et al (2017) Foraging is determinant to improve smallholders' food security in rural areas in Mali. West Africa Sustain 9:2074. https://doi.org/10.3390/su9112074

Nakamura EM, Hanazaki N (2016) Protected area establishment and its implications for local food security. Hum Ecol Rev 22:1–22

Nasi R, Taber A, Van Vliet N (2011) Empty forests, empty stomachs? Bushmeat and livelihoods in the Congo and Amazon basins. Int For Rev 13:355–368. https://doi.org/10.1505/146554811798293872

Nielsen MR, Meilby H, Smith-Hall C et al (2018) The importance of wild meat in the global south. Ecol Econ 146:696–705. https://doi.org/10.1016/j.ecolecon.2017.12.018

Nyahongo JW, Holmern T, Kaltenborn BP, Røskaft E (2009) Spatial and temporal variation in meat and fish consumption among people in the western Serengeti, Tanzania: the importance of migratory herbivores. Oryx 43:258–266. https://doi.org/10.1017/S0030605307991127

Oldekop JA, Holmes G, Harris WE, Evans KL (2016) A global assessment of the social and conservation outcomes of protected areas. Conserv Biol 30:133–141. https://doi.org/10.1111/cobi.12568

Pattiselanno F, Lubis MI (2014) Hunting at the Abun regional marine protected areas: a link between wildmeat and food security. HAYATI J Biosci 21:180–186. https://doi.org/10.4308/hjb.21.4.180

Popkin BM (2001) The nutrition transition and obesity in the developing world. J Nutr 131:871S–873S. https://doi.org/10.1093/jn/131.3.871s

Pouliot M, Treue T (2013) Rural people's reliance on forests and the non-forest environment in West Africa: evidence from Ghana and Burkina Faso. World Dev 43:180–193. https://doi.org/10.1016/j.worlddev.2012.09.010

Powell B, Thilsted SH, Ickowitz A et al (2015) Improving diets with wild and cultivated biodiversity from across the landscape. Food Secur 7:535–554. https://doi.org/10.1007/s12571-015-0466-5

Reed J, van Vianen J, Foli S et al (2017) Trees for life: the ecosystem service contribution of trees to food production and livelihoods in the tropics. For Policy Econ 84:62–71. https://doi.org/10.1016/j.forpol.2017.01.012

Rentsch D, Damon A (2013) Prices, poaching, and protein alternatives: an analysis of bushmeat consumption around Serengeti National Park, Tanzania. Ecol Econ 91:1–9. https://doi.org/10.1016/j.ecolecon.2013.03.021

Reuter KE, Randell H, Wills AR, Sewall BJ (2016) The consumption of wild meat in Madagascar: drivers, popularity and food security. Environ Conserv 43:273–283. https://doi.org/10.1017/S0376892916000059

Richardson RB (2010) Ecosystem services and food security: economic perspectives on environmental sustainability. Sustainability 2:3520–3548. https://doi.org/10.3390/su2113520

Rosendo S, Brown K, Joubert A et al (2011) A clash of values and approaches: a case study of marine protected area planning in Mozambique. Ocean Coast Manag 54:55–65. https://doi.org/10.1016/j.ocecoaman.2010.10.009

Rowland D, Ickowitz A, Powell B et al (2017) Forest foods and healthy diets: quantifying the contributions. Environ Conserv 44:102–114. https://doi.org/10.1017/S0376892916000151

Savage A, McIver L, Schubert L (2019) Review: the nexus of climate change, food and nutrition security and diet-related non-communicable diseases in Pacific Island countries and territories. Clim Dev 12:120–133. https://doi.org/10.1080/17565529.2019.1605284

Schulte-Herbrüggen B, Cowlishaw G, Homewood K, Rowcliffe JM (2013) The importance of Bushmeat in the livelihoods of West African cash-crop farmers living in a faunally-depleted landscape. PLoS One 8:1–14. https://doi.org/10.1371/journal.pone.0072807

Shafer CL (2015) Cautionary thoughts on IUCN protected area management categories V-VI. Glob Ecol Conserv 3:331–348. https://doi.org/10.1016/j.gecco.2014.12.007

Sunderland TCH (2011) Food security: why is biodiversity important? Int For Rev 13:265–274. https://doi.org/10.1505/146554811798293908

Sunderland T, Powell B, Ickowitz A et al (2013) Food security and nutrition: the role of Forest. Discuss Pap:1–8. https://doi.org/10.1016/S0306-9192(00)00026-9

Sylvester O, Segura A, Davidson-Hunt I (2016) The protection of forest biodiversity can conflict with food access for indigenous people. Conserv Soc 14:279–290. https://doi.org/10.4103/0972-4923.191157

Temsah G, Johnson K, Evans T, Adams DK (2018) Benefits of biodiverse marine resources to child nutrition in differing developmental contexts in Hispaniola. PLoS One 13:1–24. https://doi.org/10.1371/journal.pone.0197155

Timko JA, Satterfield T (2008) Seeking social equity in National Parks: experiments with evaluation in Canada and South Africa. Conserv Soc 6:238–254. https://doi.org/10.4103/0972-4923.49216

Turner NJ, Turner KL (2008) "Where our women used to get the food": cumulative effects and loss of ethnobotanical knowledge and practice; case study from coastal British Columbia. Botany 86:103–115. https://doi.org/10.1139/B07-020

Turner NJ, Luczaj LJ, Migliorini P et al (2011) Edible and tended wild plants, traditional ecological knowledge and agroecology. CRC Crit Rev Plant Sci 30:198–225. https://doi.org/10.1080/07352689.2011.554492

Van Vliet N, Nebesse C, Nasi R (2015) Bushmeat consumption among rural and urban children from province Orientale, Democratic Republic of Congo. Oryx 49:165–174. https://doi.org/10.1017/S0030605313000549

Vinceti B, Termote C, Ickowitz A et al (2013) The contribution of forests and trees to sustainable diets. Sustain 5:4797–4824. https://doi.org/10.3390/su5114797

Walsh-Dilley M, Wolford W, McCarthy J (2016) Rights for resilience: food sovereignty, power, and resilience in development practice. Ecol Soc 21. https://doi.org/10.5751/ES-07981-210111

Weerasekara PC, Withanachchi CR, Ginigaddara GAS, Ploeger A (2018) Nutrition transition and traditional food cultural changes in Sri Lanka during colonization and post-colonization. Foods 7:1–18. https://doi.org/10.3390/foods7070111

West P, Igoe J, Brockington D (2006) Parks and peoples: the social impact of protected areas. Annu Rev Anthropol 35:251–277. https://doi.org/10.1146/annurev.anthro.35.081705.123308

Worm B, Barbier EB, Beaumont N et al (2006) Impacts of biodiversity loss on ocean ecosystem services. Science (80) 314:787–790. https://doi.org/10.1126/science.1132294

Wunder S, Angelsen A, Belcher B (2014) Forests, livelihoods, and conservation: broadening the Empirical Base. World Dev 64:S1–S11. https://doi.org/10.1016/j.worlddev.2014.03.007

Young H, Taylor A, Way S-A, Leaning J (2004) Linking rights and standards: the process of developing "Rights based" minimum standards on food security, nutrition and food aid. Disasters 28:142–159. https://doi.org/10.1111/j.0361-3666.2004.00249.x

Zavaleta C, Berrang-Ford L, Llanos-Cuentas A et al (2017) Indigenous Shawi communities and national food security support: right direction, but not enough. Food Policy 73:75–87. https://doi.org/10.1016/j.foodpol.2017.10.001

Chapter 5
Challenges for Rural Livelihoods, Participatory Agroforestry, and Biodiversity Conservation in a Neotropical Biosphere Reserve in Mexico

Luis García-Barrios, Juana Cruz-Morales, Marco Braasch, Yanus Dechnik-Vázquez, Alonso Gutiérrez-Navarro, Amayrani Meza-Jiménez, Tlacaelel Rivera-Núñez, Erika Speelman, Gabriela Trujillo-Díaz, Vivian Valencia, and Aiora Zabala

5.1 Introduction

The *Cuenca Alta del Río el Tablón* (CART), or Upper Watershed of the Tablón River, is located in the northwestern portion of Chiapas's Sierra Madre. Between 1960 and 1990, most of the land in the CART was assigned to *ejidos* (rural communities with collective tenure rights and social organization). In 1995, it was decreed as part of the *La Sepultura* Biosphere Reserve (REBISE). The CART includes both the Sepultura's nucleus (known as *Tres Picos*) and a significant portion

L. García-Barrios (✉) · Y. Dechnik-Vázquez · A. Meza-Jiménez · T. Rivera-Núñez
El Colegio de la Frontera Sur (ECOSUR), San Cristóbal de las Casas, Chiapas, Mexico

J. Cruz-Morales · G. Trujillo-Díaz
Universidad Autónoma de Chapingo Campus Chiapas,
San Cristóbal de las Casas, Chiapas, Mexico

M. Braasch
Forstbetrieb Fürst zu Fürstenberg, Donaueschingen, Germany

A. Gutiérrez-Navarro
Universidad Nacional Autónoma de México, Ciudad de México, Mexico

E. Speelman
Wageningen University & Research, Wageningen, The Netherlands

Wageningen University, Wageningen, The Netherlands

V. Valencia
Wageningen University & Research, Wageningen, The Netherlands

A. Zabala
University of Cambridge, Cambridge, UK

C. Baldauf (ed.), *Participatory Biodiversity Conservation*,
https://doi.org/10.1007/978-3-030-41686-7_5

of its buffer zone (Valdivieso-Pérez et al. 2009). Its altitudes range from 800 to 2500 m above sea level, and its climates from warm and dry to temperate and humid, with a seasonal dry spell that affects the whole region. The CART's homogeneous granitic lithology generates coarse, sandy soils with variable (but generally high) susceptibility to erosion, and moderate to high fertility (Valdivieso-Pérez et al. 2009; Márquez-Conrado and Cruz-Morales 2004). It has a dense network of streams (most of them seasonal) that are tributaries of the *El Tablón* River. The watershed was originally covered by deciduous forest in its lower reaches; by forests of pine, oak, and riparian sub-perennials in the middle part; and by mountainous cloud forest and scrublands in the upper part. In general, the degrees of deforestation and agricultural transformation decrease as one ascends the mountain range, and are virtually zero in the nucleus. Populations of birds, mammals, reptiles, amphibians, and arthropods appear significantly reduced not only in the open areas and woodlands of the buffer zone, but also in many parts of the nucleus zone.

More than 90% of the land that is included in the nucleus zone belongs to *ejidos*, and the rest to smallholders. The total area of the 16 *ejidos*, founded in the mountainous parts between 1960 and 1990, is 23,000 ha. The approximately 6000 inhabitants of those *ejidos* are distributed among almost 1500 households: 25% of the heads of families have full land rights (*ejido* members), and the other 75% are either descendants who have acquired land from *ejido* members (town dwellers), or are residents of the *ejidos* without access to land (nonmember residents; IMSS 2014). Almost all of the *ejidos* have between 1000 and 2500 ha, but some have less than 500 ha. The amount of land per household is, frequently, unequal within *ejidos* as well as among them.

The majority of the population that arrived to colonize the Sierra were mestizos from the continuous region, known as the *Frailesca*; a minority were indigenous from the Chiapas Highlands (Tseltales, Totsiles, Chichonales, or Zoques). Within a short period of time—which began around 1960—these agrarian colonists and their descendants have had to construct and reconstruct repeatedly their strategies for social reproduction in response to the changing interactions among multiple actors, internal and external. Quite frequently, external actors were more powerful than the colonists, and worked in many social venues to dispute both the territories that the colonists were constructing, and the benefits derived therefrom. Throughout the last 60 years, this continuous and sometimes dramatic changing of strategies has occurred within a watershed that is ecologically heterogeneous, and has been accompanied by population growth and great differences in the opportunities and capacities of groups to meet their needs and create livelihoods. These differences have generated an enormous diversity of conditions and social positions among rural domestic groups (RDGs) and, consequently, a great diversity of ways in which agriculture is practiced and contributes to social reproduction.

The population of each *ejido* is concentrated in a single settlement, rather than being dispersed. These small population centers are connected by dirt roads that are not easily traveled. Almost all of the houses have concrete floors, piped water, and electricity, but their size and quality is notoriously variable. Before 2017, about half were of adobe with tile roofs, and had a small living room (*sala*), a kitchen with semi-open cooking stove, and one or two bedrooms. The adobe houses were severely damaged by the strong earthquake of September 2017, and are now being replaced, when possible, by houses made of concrete block with galvanized-steel roofs.

At present, the gradient of land uses within the watershed's dense network of narrow valleys and steep hillsides is diffuse, but perceptible. The gradient is primarily due to changes in altitude and steepness of slopes, and secondarily to differences in sizes of properties and the economic strategies of owners. RDGs with abundant lands, at lower elevations, raise cattle and grow primarily corn, beans, and sorghum. Those with little land, in the higher regions contiguous with the nucleus, grow shade coffee and the ornamental palm tree, *Palma Camedor* (*Chamaedorea* spp.), plus some corn. RDGs with medium-sized lands, or at medium elevations, may do any of the above. Almost all RDGs have access to wood of diverse types for domestic use, including oak for firewood; some RDGs extract pine lumber or resin (Braasch et al. 2017). It is important to note that the CART's network of streams is notoriously underutilized for irrigating crops. Currently, the majority of forestry and agricultural production goes to regional markets, plus some international ones. Producers of corn and beans grow those crops for their own use, and sell surplus.

Only the CART's largest population center generates opportunities for self-employment in small businesses and in the non-agricultural sector, and provides these services to the entire watershed. For many RDGs, a very significant part of family income derives from seasonal work in the US and northern Mexico, plus payments from government programs. Slightly more than half of the RDGs are pluri-active (that is, they have multiple sources of income from non-agricultural sources as well as agricultural ones). A minority of RDGs specializes in some form of agricultural production, and an increasing number depends upon government payments and migradollars (remittances from family members who have migrated to the US).

RDGs within a given *ejido* form networks of persons linked by marriage, ritual practices, and blood relationship. These networks act on private and public levels through the *ejido's* assembly; in groups that solicit and administer government aid; in the weakened Catholic congregations; and throughout the variety of growing evangelical churches. The organizations for marketing agricultural products are weak and volatile, and are oriented more toward receiving and distributing subsidies than toward collective actions for commercializing products and negotiating better terms to obtain inputs. The relationship with the Reserve's administrators has been limited, clientelist, and complicated. Tensions among town dwellers, *ejido* members, and nonmember residents of *ejidos* are strong, and the gap between interests of youths and adults is widening. Fortunately, this region of the Sierra has not seen episodes of physical violence associated with the presence of organized crime, but discrete consumption of drugs by young people is increasing. The majority of the population lives between hopes of a better life through unceasing economic effort and an unbreakable faith in God, and the uncertainty provoked by rapidly changing environmental, economic, and social conditions that challenge residents' abilities to maintain that which has already been achieved.

5.2 Construction of Territories by *Campesinos* in the CART Before the Neoliberal Structural Reforms (1960–1995)

Until 1960, most of the CART was classified formally as "national lands" (under the ownership of the Mexican government), but its resources in the higher altitudes were used intermittently by harvesters of *Palma camedor* and by medium-sized sawmills (illegal, but tolerated). At lower elevations, tracts of forest and meadow were appropriated *de facto* for transhumant cattle-raising by ranchers from adjoining valleys in the *Frailesca* and from the capital city of the state of Chiapas, Tuxtla. To a much lesser extent, ranchers allowed their laborers and foremen to use the same lands for share-cropped production of corn, to meet the needs of the workforce.

During the period from 1940 to 1980, there were 59 ranches, owned by 22 families, occupying slightly more than 22,000 ha. The federal land-redistribution program, which granted *ejidos* official recognition, arrived late in Chiapas and especially in the Sierra, because of the anti-agrarian *mapachista* movement.

Marginal agro-silvo-pastoral activity in the CART dates back to 1860, but between 1960 and 1995 the founding *ejidatarios* (*ejido* members, many of them former landless wage laborers or *peones* on ranches) followed a new land-use strategy that was clearly oriented toward colonization, appropriation of space, and rapid expansion of the agricultural frontier over thousands of acres of forest, with the help of fire. In their own words: "It was exciting to see the mountains burning at night." Through arduous labor, the population constructed a *campesinos* territory, which was agricultural and *ejido*-based, and whose fundamental economic and political driver was, until the 1980s, the incorporation of the Sierra into the boom of commercial, high-input corn production that was already underway in the valleys of the contiguous *Frailesca* (one of Mexico's three granaries).

In the style of the ranchers who had been their masters, the *ejido* populations maintained, as secondary market activities, the same transhumant cattle-raising in the forests, along with lumbering and extraction of *Palma Camedor*. The former *peones* also constructed their own coffee farms under the diverse shades of native and cultivated trees.

After taking legal possession of the land, social reproduction of RDGs in *ejidos* occurred under the guardianship of a benefactor State, corporate and authoritarian, which demanded loyalty and party discipline from this population, while also offering agricultural credit, crop insurance, some types of medicines, and guaranteed prices for its products. This guardianship did not raise most of the population out of poverty, nor did it provide all of the required services and infrastructure, or even impede (in any degree) the voracity of merchants, loan sharks, and land speculators. However, it did provide some stability to the process of social reproduction of RDGs in the Sierra.

5.3 Social Reproduction of Domestic Groups in the CART During the Period of Neoliberal Globalization (1995–2017)

During the period 1950–1985, Mexico's rural population had the task of providing food and labor at low cost to maintain salaries at the low levels required by the process of capitalist urban-industrial growth. Agrarianism and paternalistic guardianship were oriented toward guaranteeing that function. The State's investment in the peasantry was insufficient and uneven: Chiapas—especially its mountainous regions—was relegated during most of the period, thereby leaving plenty of territory to its rural oligarchies and local bossdoms. The campesino regions of southern Mexico lagged behind in the country's capitalist programs, to which then the effects of the predominant neoliberalism of the last 30 years were added. The neoliberal processes most relevant to our case were: (1) a halt in both agricultural land distribution and legalizing the privatization of *ejido* properties (including the sale thereof to individuals); (2) putting national lands and *ejidos* forests under government protection as National Protected Areas, after which some have been opened gradually to corporate exploitation of minerals, aquifers, organisms, and scenic values; (3) massive importation of cheap basic grains for urban and rural consumption, and orientation of aid received by rural regions to the production of agricultural goods for export to markets in North America and Europe; (4) dismantling of the government's direct role in providing technical training and credit to *campesinos*, and also eliminating the state's role in regulating terms of interchange and wealth distribution in rural chains of production and supply; (5) channeling the landless and unemployed of rural areas into seasonal and permanent employment in the north of Mexico and in the US; and (6) managing the poverty of the non-migrating rural population through subsistence government payments, and through small subsidies for precarious production of food for self-consumption.

How have these processes manifested themselves in the CART? In the mid-1990s, various events destabilized the social-reproduction strategies of the majority of RDGs. *Ejidos'* requests for additional land were canceled, and the program *Programa de Certificación de Derechos Ejidales* (PROCEDE) was initiated to give private titles to *ejido* members, thereby legalizing the purchase and sale of land. The CART was declared, in its entirety, to be part of *La Sepultura* Biosphere Reserve. The nucleus zone included the national lands (plus some private properties and sections of *ejidos*), while the buffer zone included most of the remaining *eijdos*, and was subject to further zoning and a management plan generated with minimum local participation. New regulations prohibited further deforestation as well as hunting and harvesting of timber, wood products, and ornamental palm. The use of fire as part of agricultural practices was also outlawed. Finally, profitability of corn production plummeted due to massive importations under the North American Free Trade Agreement, and because the Mexican government canceled subsidies for agrochemicals.

Through an enormous effort, in open or secret resistance to and confrontation with the *Comisión Nacional de Áreas Naturales Protegidas* (CONANP, the National Comission of Protected Natural Areas), not devoid of personal suffering, and with no alternative, the RDGs in lower- and middle-elevation zones of the CART reorganized their strategies of reproduction. Using dollars sent by waves of migrants, plus modest subsidy payments from *la Secretaría de Agricultura y Ganadería* (Secretariat of Agriculture and Animal Husbandry), part of the population found a way of adapting to the new circumstances: they bought cattle, extended grazing lands over old cornfields and forest edges, and grazed cattle more intensively within the forests. The RDGs who adopted this strategy transitioned from supplying corn for the national market to depending upon migration, and thereby to linking themselves with the lowest and least profitable level of the value chain: the production of livestock for exportation, via the sale of weaned calves.

The cattle-raising strategy of these RDGs soon encountered resistance from outside. The unprecedented forest fires of 1998, associated with a severe *El Niño* phenomenon, and with the expansion of exotic grasslands in the REBISE, motivated the CONANP, the Comisión Nacional Forestal (CONAFOR, the National Forestry Comission), and international conservation groups—The Nature Conservancy (TNC) and Conservation International (CI)—to demonize extensive cattle ranching. As a result, in 2004, the program of payments for hydrologic services was instituted by CONAFOR, which aimed at preventing and controlling fires by implementing intensive measures. Those measures would be operated by CONANP/CONAFOR and *ejidal* assemblies. The number and extent of fires have since decreased considerably (Gutiérrez-Navarro 2015). Although many RDGs appreciate the benefits, they came at the cost of a population explosion of ticks and rodents, the aging of forage plants in pastures, and increased use of insecticides and herbicides. After 2006, CONANP accepted that cattle-raising in the CART would not stop, and instead became interested in proposals to cultivate forage trees and fodder crops (for cutting and subsequent storage or immediate use) so that cattle could be concentrated at lower altitudes, thereby reducing browsing in the forests. Therefore, CONANP appointed a promoter of sustainable cattle-raising for each *ejido*. The program, which had modest material support, quickly found itself reduced to a small group of cattle-raisers who produced milk and cheese, the majority of whom have adopted only fodder crops, and have difficulty working together. They have been reluctant to suspend browsing in the forests (as would be necessary in order to meet CONANP's requirement that "forests be freed for restoration"; Box 5.1).

Box 5.1 Silvopastoralism in the CART: Bring the Cow to the Forest, or Bring the Forest to the Cow?

Efforts to promote silvopastoral practices are important for conservation within the REBISE, but have had varying results. Some ranchers adopt intensifying practices labeled silvopastoral, yet are firmly grounded in cultivation of *Pennisetum* "elephant" grasses. In the cases where trees are used, these are kept cut at low height for cattle to reach the foliage, and thus actual mature trees do not grow. Admittedly, most ranchers still have extended pastures and more conventional intensification practices as the backbone of their ranching activities.

Is there a place for the forest then, in such panorama? As it turns out, there *is* a firmly grounded use of the forest by most of the lower-resource smallholders. In a subtropical region like the CART, the dry season limits the capacity to feed cattle. Having a parcel or terrain with forest allows for liberating cows into it, "to let them find what they can," when there are no other resources available. Others, not having such needs, still let their cattle roam into forested areas, assuring that browsing augments the quality of meat/milk. Bypassing judgments about conservation, we can see that smallholders *are actually using the forest frontiers as silvopastoral systems.*

We did not know exactly, though, to what extent the forest frontiers were rife with browsing species. We thus surveyed the richness of these landscapes. Results showed high numbers of browsed plants: 161 of 268 total species. Oak forest had 71 foraged plants (58% of the species within this community); riparian vegetation and tropical dry forest had 72 (47%) and 88 (54%) each. Some of these have much potential for silvopastoral intensification (Dechnik-Vázquez et al. 2019). In such a landscape, it makes sense why smallholders (both in need and not) value browsing of the forest, being reluctant to abandon this practice. These landscapes in forest frontiers are direct indications of how to further design silvopastoral systems that better mimic what we see in the forest; maybe by plans that involve directed succession.

Seeing the situation through a different lens, all this actually suggests that a more intimate symbiosis of ranching and conservation can be achieved. We have to bypass some traditional judgments on the use of the forest by cattle, and develop plans that continue efforts to conserve patches of forest (and plantations with actual grown trees), and admit some silvopastoral use of the forest frontiers themselves.

Elaborated by Yanus Dechnik-Vázquez

In the semi-open pine forests at middle altitudes—previously subject to grazing, burning of meadows, and lumbering—the conflict between the RDGs and external actors was particularly intense because CONANP's prohibitions and restrictions on those practices left the RDGs with almost no alternative options. On their own initiative, and with the support of outside actors, in 2005 the *ejidatarios* began to explore the possibility of extracting and selling pine resin. Five years later, in 2010, the Secretariat of Environment and Natural Resources (SEMARNAT) authorized the extraction of resources, and in 2012 signed a 10-year contract for the sale of resin to the corporation *AlEn del Norte, S.A. de C.V.* Thus began the resin-extraction activity, which has since provided a modest but dependable income to participants during the dry season. Because of differences within groups of resin extractors, and because of a combination of low resin prices and high costs of certification, approximately 50% of the project's original partners have suspended this activity (Box 5.2).

Box 5.2 Resin Harvesting: A Productive Activity that Contributes to Biodiversity Conservation and Farmers' Livelihood

Over decades, smallholder farmers living at mid-altitudes within the CART have transformed closed oak–pine forests into semi-open pine forests. Oak forest stands in valley-bottoms were deforested for agriculture and fire was used to prepare those terrains for cultivation. These burnings frequently triggered forest fires and led to high mortality of saplings and adult trees. In steeper areas, farmers extracted pine trees for construction and oak trees for fuel wood use. In the late 1990s, livestock production increased and cattle-raising in the forest became common (Braasch et al. 2017).

In 1995, when the REBISE was declared, CONANP promoted reducing cattle production inside and near forests, regulated commercial wood extraction, and forbid the use of fire on agricultural land within forested areas. This restrictive conservation management led to conflicts between farmers and CONANP.

In 2005, *ejido* members started to explore the possibility of extracting and selling the resin of *Pinus oocarpa*. Farmers, supported by external actors, visited the state of Michoacán, México, where for more than 80 years resin harvesting has been practiced. This knowledge exchange between farmers increased the interest of a resin project in the CART, which was finally authorized by SEMARNAT in 2010. In 2012, a contract was signed with the resin corporation *AlEn del Norte*, and ever since the resin revenue has contributed significantly to the farmers' livelihood. The main benefit of resin harvesting, compared to other economic activities, is the potential to produce all year round with a peak during the dry season, when wage work is usually scarce. Because of the economic benefits of resin harvesting, farmers have avoided the use of fire and increased their interest in protecting and reforesting pine trees.

Resin-producing trees need to be replaced after approximately 20 years of extraction. However, because of the invasion of exotic grasses into the forest and fire suppression management, grass biomass that inhibits tree regeneration has accumulated. To reduce the dense grass cover and encourage pine regeneration, farmers and scientists suggest cattle grazing. However, this is not accepted by all actors, because cattle trampling may cause soil degradation, and increases seedling and sapling mortality. To understand the trade-offs between exotic grass expansion, pine regeneration, and cattle grazing, we developed the game "TRUE-GRASP." The game was applied in workshops to provide a space for collective learning among actors, to solve conflicts, and to negotiate agreements, in order to facilitate resin harvesting as a long-term productive activity that contributes to biodiversity conservation and farmers' livelihood (Braasch et al. 2018).

Elaborated by Marco Braasch

Beginning in 2014, the prices of calves and corn increased considerably, but were very volatile. In the CART, the response to the price fluctuations has been a visible and frequent rotation between pastureland and cornfields. However, extensive cattle-raising continues to be the preponderant source of income for those RDGs that have the largest land holdings in the CART's lower- and middle-elevation portions. RDGs with little or no land dedicate themselves to corn production, and depend to a greater extent upon government transfers and remittances from migrants. In general, the majority of RDGs in these two zones have approximately 20 ha of land and are pluri-active. Due to the scale or intensity (or both) of their economic operations and levels of investment, almost none of the RDGs can be classified as family businesses.

For RDGs in the highest-elevation regions of the CART, next to the nucleus and therefore monitored most closely by the CONANP, the initial changes during this period were particularly damaging. The *ejidos* with the least land, and those dedicated more to harvesting *Palma camedor* than to raising cattle and corn, were left without any alternative options. Thus, they resorted to migration of men and women, and to wage labor in other *ejidos*, while simultaneously evading and resisting the decisions of environmental authorities as best they could. Upper-elevation RDGs who owned cattle and cultivated corn experienced dynamics similar to those in the middle- and low-altitude zones. One village was dismantled, and two others were unable to complete the process for becoming *ejidos* and remained as irregular settlements. To help formulate agricultural projects that would meet the interests of the *ejidatarios* involved, various external non-governmental actors mediated and provided training. From those efforts arose the controlled extraction of commercial lumber, payment for environmental services linked to control of fire, production of shade-grown coffee certified as organic, cultivation of the ornamental palm *Chamaedorea quezalteca* in the understory, and raising of ornamental cycads in greenhouses. Some *ejidos* engaged in all of these activities, but the majority engaged in only one. The CONANP's strategy was to link these projects/programs to international markets, which proved very volatile. In 2014, after the international corporation Floral Green suspended purchasing, the production of palm entered into a crisis, and it took years to find alternative national customers. Coffee producers never completely benefited from the organic seal because the purchase of certified coffee was monopolized by *Agroindustrias Unidas de México* (AMSA), in agreement with CI and Starbucks. Beginning in 2013, the coffee rust *Hemileia vastatrix* devastated plantings, reducing production by 40–50% and opening the doors to new private-sector actors (agroexporters, greenhouse owners, and banks) to promote contracts for reestablishing coffee plantings with resistant hybrid varieties, which were promoted as requiring full sun exposure (Valencia et al. 2018). Still, the CONANP and many producers are interested in growing the new varieties under shade (Box 5.3). In times of production collapse, the RDGs in higher elevations have depended strongly upon combinations of remittances from migrants in the north of Mexico, government transfers, and small payments for environmental services. The majority of income from those sources is spent on housing and industrialized foods.

Box 5.3 New Challenges for Biodiversity Conservation and Livelihoods in Agroforestry Systems

Coffee agroforestry systems are a promising alternative to conventional agriculture that may address the dual goals of biodiversity conservation and development. However, how farmers manage forests for coffee cultivation may affect the potential of agroforestry systems to conserve biodiversity. A decade of research in *La Sepultura* Biosphere Reserve has improved our understanding of why and how farmers manage agroforests the way they do, the outcomes on tree biodiversity conservation, and how shocks (e.g., coffee leaf rust) may disrupt an entire socio-ecological coffee-growing landscape. Our research has shown that although tree biodiversity is similar between coffee agroforests and forest, tree community composition is significantly distinct: coffee agroforests have higher abundances of pioneers, lower abundances of trees of conservation concern, and are dominated by farmers' favorite trees, particularly *Inga* spp.

Coffee leaf rust has acted as a shock that reconfigured the socio-ecological landscape, including coffee-growing practices. Before the coffee leaf rust outbreak, most farmers cultivated *Coffea arabica* (Arabica) varieties and were certified organic. Farmers repeatedly reported their disappointment with the small differential between organic and conventional farmgate prices. The low premiums received by farmers were mostly the result of price deductions due to certification costs and penalizations from buyers due to quality. After coffee leaf rust outbreak, farmers started applying fertilizers and pesticides in an unsuccessful attempt to control the pest. Farmers did not seem concerned with losing their certification status; it is uncertain whether farmers will attempt to reenter certification schemes.

In response to coffee leaf rust, farmers have adopted new hybrid coffee varieties that are currently resistant to the pathogen. New hybrid coffee varieties were introduced by external agents and described as sun tolerant, and suitable for lower elevations and dryer and warmer climates as compared to the Arabica varieties traditionally grown in the region. Since about 2015–2016, farmers have gradually adopted these new hybrid varieties either by substituting affected Arabica coffee bushes in their coffee agroforests, or by setting new coffee fields with sparse shade trees at lower elevations in substitution of pastureland or corn and bean fields. Whether these land-use and land-cover change trends will continue in response to coffee leaf rust is unclear. *La Sepultura* Biosphere Reserve, along with other coffee-growing landscapes in Mesoamerica, is undergoing a socio-ecological reconfiguration process in response to coffee leaf rust.

Elaborated by Vivian Valencia

The period 1995–2017 in the CART saw two distinct groups of social and territorial actors engaging each other in a process of creating, resisting, and negotiating the real makeup of the REBISE. The first group included inhabitants of the REBISE (territorial actors). The second group was formed by extraterritorial actors such as governmental institutions and non-governmental organizations (NGOs), who, although external to the REBISE, intervened therein, generally by means of public policies, programs, and other actions. The territorial and extraterritorial social actors who lived and worked face to face configure a space of social relations, defined to a great extent by public policies, socio-environmental situations, and the macro- and microeconomics described above.

During the last 23 years, managers of the REBISE have undertaken two major tasks to construct a conservation territory in the CART: (1) prohibiting activities and designing management plans for the watershed by using the meager resources that they administer directly, and by establishing clientelist relationships with small groups in each *ejido*, with which the managers of REBISE negotiate without ceding authority; and (2) exploring different activities to induce changes in land use that are (or are believed to be) favorable to the conservation of forests and local lifestyles. The short- and medium-term projects promoted by the managers include: (a) modest actions for reforestation, and for restoration of areas previously cultivated, in order to stimulate carbon sequestration; (b) prevention of fires through payments for hydrological environmental services; (c) production and commercialization of coffee, *Palma camedor*, ornamental cycads, and pine resin; (d) use of alternative fodders to eliminate grazing of the forest understory by cattle; (e) backyard activities (carried out by groups of women) such as making compost and raising fruits, vegetables, and fowl; (f) firewood-saving stoves; (g) campesino schools that provide workshops about coffee, ornamental palm, silvopastoral cattle-raising, and integral management of fire; (h) diagnoses and studies commissioned by NGOs and universities (diagnoses of communities in the first years, and later diagnoses per product); (i) vertical direction by the Advisory Council of the REBISE; (j) promotion of a network of assistance for the REBISE, consisting of governmental and academic institutions and NGOs (2005–2007); (k) participation at the district and municipal levels in the Council for Sustainable Rural Development (2005–2007); (l) annual participation in the National Conservation Week; and (m) occasional participation in *ejido* assemblies, and in meetings with other extraterritorial actors. These activities make clear that local and extraterritorial actors have had many needs and opportunities for interacting with each other in the zone. Unfortunately, the processes that these modest actions might have triggered for enabling the population to find logic and sustenance in the ecological territory of the CART, beyond what is considered indispensable for guaranteeing their short- and medium-term social reproduction, have been thwarted by the combination of the meager (and decreasing) human and operational resources assigned to the REBISE, the vertical and clientelist inertia of relations among actors, the low level of trust and participation of the local population in these projects, the economic polarization of local RDGs, the volatility and unfavorable condition of "green markets," and the State abandonment of campesino agriculture (see complete explanation in Cruz-Morales 2014a, b; Cruz-Morales and García-Barrios 2017).

We have acted in this field of social interaction as researchers and postgraduate students of the *Centros Regionales de la Universidad Autónoma Chapingo* (2002–2018) and *El Colegio de la Frontera Sur* (ECOSUR; 2006–2018). It took time to find our place in what has shown itself to be a multifaceted, complex, and changing reality, and to insert ourselves usefully in the relations of cooperation, competition, and conflict that occur within the *ejidos*, and between the *ejidos* and the CONANP whenever the population makes use of the agricultural and natural heritage that it possesses in the CART. To maintain a long-term presence in the territory, we established bonds of trust and showed ourselves to be dependable, frank, and transparent in our actions. We also demonstrated our commitment by responding to invitations to events held by the communities and the CONANP. We developed a multidimensional research effort by constructing a multidisciplinary and interinstitutional network of collaboration. Research efforts have operated with a minimal budget but with the enthusiastic endeavor of a small, permanent group of personnel from ECOSUR and the *Universidad Autónoma de Chapingo*, as well as 30 graduate students and their academic advisors from various universities in Mexico, the United States, Puerto Rico, the United Kingdom, France, the Netherlands, Germany, and Italy.

In the process, it has been crucial to explore, undogmatically, diverse modes of participatory-action research appropriate to each situation and objective, in order to become familiar with the ecological, agricultural, economic, social, and political processes within the CART. We also needed to understand the different actors, and to reflect with them openly, critically, and constructively. At all times, we have been privileged to understand and respond to the RDGs' own perspectives on social reproduction. Our academic centers have limited operational and financial capacity, and their independent, critical approaches to difficult social and environmental realities find few receptive ears among other external actors. For that reason, our work has been oriented, in most cases, toward diverse pilot experiences with small groups of interested local and external actors. Some of those experiences have had very modest local impact, while others have borne fruit after many years.

This participatory-action research, during the period 2004–2017, contributed to the following:

1. Sensitizing the CONANP and conservationist NGOs to the present and historical conditions of social reproduction of the population established in the CART (their identities, necessities, restrictions, motivations, interests, etc.), and to the need to construct rather than impede direct democracy and environmental citizenship in this *campesino* territory (Márquez-Conrado and Cruz-Morales 2004; Cruz-Morales 2014a, b).
2. Explaining the processes and driving forces that have led to frequent changes in the appropriation and use of land (Márquez-Conrado and Cruz-Morales 2004; Valdivieso-Pérez et al. 2009; Speelman et al. 2014a, b; Zabala 2015), and the consequences for floristic composition (Valencia et al. 2014, 2015, 2016), vegetative cover (Sanfiorenzo-Barnhard et al. 2009; Sanfiorenzo-Barnhard 2012; Valencia et al. 2018), and soil quality (Valdivieso-Pérez et al. 2012).

3. Revealing the present low level of consensus among and within groups of local and external actors regarding which territory's resources are most important, and which problems take priority (Brunel-Manse and García-Barrios 2011).
4. Comprehending in depth the forms of cattle-raising practices, the obstacles faced, and the place of silvopastoral practices in people's search for solutions (García-Barrios et al. 2012; Rosabal-Ayan 2015; Dechnik-Vázquez et al. 2019).
5. Constructing capacities and consensus as to which land-use and cattle-raising practices are best in their area, based on local and academic knowledge (Cruz-Morales et al. 2011). This effort contributed to CONANP's change in opinion and actions regarding cattle-raising in the territory.
6. Motivating 200 cattle-raising *campesinos* to establish in their premises small experimental modules with forage trees and other species (García-Barrios et al. 2012; Cruz-Morales et al. 2011; García-Barrios and González-Espinosa 2017).
7. Analyzing the forms in which producers participate in field experiments, and studying producers' opinions about silvopastoral innovation, thereby revealing the ecological limitations and the social obstacles and opportunities that present themselves when innovative cattle-raising practices are tested (Zabala et al. 2017).
8. Experimentally developing agroecological practices that might reduce technical and economic restrictions identified by cattle-raisers as impediments to making silvopastoral innovation attractive and successful (Vides-Borrell et al. 2011; Morales-Díaz 2011; Oleta-Barrios 2012; Buhmann and García-Barrios 2014; Braasch et al. 2018; Dechnik-Vázquez et al. 2019).
9. Diagnosing the presence of improperly disposed solid waste in the fluvial network of five *ejidos* (Cruz-Morales et al. 2015); promoting the collection, alternative use, and sale of polyethylene terephthalate (PET) plastic refuse in two *ejidos*; and systematizing social perceptions regarding solid waste in an *ejido* (Figueroa-Sánchez 2018).
10. Generating social capacities in a population center designated as irregular, in order to support that center's attempts to constitute itself as an *ejido* and construct its territory in less vulnerable conditions (by supporting agricultural consulting, community assessments, growing fruit, educational workshops; Trujillo-Díaz et al. 2018; Cruz-Morales et al. 2014a).
11. Involving local and external actors (irrespective of age and gender) in joint construction of medium-term popular education. The emphasis has been on designing and using ten ecological and socio-ecological games (computer games as well as board games) that provide the local population with the elements needed for innovation and for adapting their community norms, agricultural practices, and modes of life (semi-specialized in cattle-raising, resin, and coffee) to new ecological and social challenges (García-Barrios et al. 2011, 2015, 2017; Speelman et al. 2014a, b; Meza-Jiménez and García-Barrios 2015; Meza-Jiménez et al. 2016; Braasch et al. 2018).

In 2015, ECOSUR approved and funded a 4-year multidisciplinary project, in which the reference frame for theorizing and carrying our practical actions was not agriculture as an abstract notion, but rather the social reproduction of domestic groups of *campesinos* within "fields" (in the sense that Pierre Bourdieu uses that term), defined to a great degree by globalized neoliberal regimes, as those regimes operate in different rural territories of Mexico's southern border region (Box 5.4). The CART, being one of the *campesino* territories considered in that 4-year project, has seen the deployment of all the methods and instruments generated collectively by researchers and graduate students associated with that project, and by other investigations carried out in the watershed by the *Family Agriculture* (AGFAM) subgroup of ECOSUR, with the most permanent involvement therein.

Box 5.4 Peasant Social Reproduction in the Forest Frontiers of the CART

The analytical framework of social reproduction was elaborated by Pierre Bourdieu (1990), to elucidate how social orders are perpetuated based on the relations (objective and subjective) that constitute them. For Bourdieu, the internal principle of the functioning of the social world depends on class fractions that are capable to generate strategies of reproduction, depending on different types and distributions of capital: economic, cultural, social, and symbolic. This is done to guarantee their conditions of life, and to try to improve their position in relation to other groups. In this "reproduction game," the different agents and/or institutions with shared interests constitute dynamic social fields, generating disputes or synergies in order to preserve privileges and prestige, and/or transform domination and social differences.

Thus, social reproduction (SR) represents a macrosocial or far-reaching theory that, by itself, does not have the capability to dimension some singularities that characterize the peasant socio-economy. It is considered that this gap can be accurately addressed by integrating Alexander Chayanov's theory of the peasant economy (1966), which recognizes domestic groups as the basic units to make decisions or generate balances, based on their working capacity. This working capacity is based on the internal morphology (number and gender of members), family development cycle (expansion, dispersion, or replacement), their agricultural–natural heritage, and political and mercantile relationships.

Given this theoretical conjugation, we developed a study with the aim of knowing the scopes of the efforts developed by external territorial agents to improve the livelihoods in the forest frontiers, based on the implementation of "green economy" type agroforestry projects. For that purpose, we analyzed the social reproduction of peasant households in *ejidos* of the upper part of the CART, specialized in agroforestry activities, and contrasted this scenario with the reproduction of households in *ejidos* in the lower part, specialized in agricultural activities. Four key results were found: (1) there is an important distinction between the disposition of capitals and heritages among peasant households; as a result, the implementation of agroforestry projects—in which each household compromises their own capitals and patrimonies to join these collective endeavors —has reinforced social polarization within the *ejidos*; (2) in the *ejidos* with agroforestry projects, households develop an average of four work activities, while the *ejidos* based on agriculture generally develop only two activities. This contrasts with the logic of external actors, who seek to promote labor specialization toward agroforestry projects, and narrow the agricultural gap in order to stabilize the forest frontiers; (3) the *ejidos* that develop agroforestry have generated interesting local dynamics of labor transfer, through the payment of rural wages by the households that have vast means of production but do not have enough labor force, and vice versa. This labor dynamic is starting to slightly reduce the strong waves of international migration, also mitigating the high dependence on government monetary transfers—the so-called remittance and subsidies landscapes, increasingly common in the rural areas of Mexico; and (4) after barely a decade of promoting agroforestry projects, and beyond the scopes of the social reproduction of households, these projects have been internalized in the imaginaries and subjectivities of the *ejidos*; as a result, it is possible to find nowadays peasant communities with resin, coffee, and palm tree exploitation *habitus*.

In conclusion, we reflected on the importance of knowing, prior to the development of any efforts for biodiversity conservation and implementation of agroforestry and silvopastoral projects, the social reproduction and economy of peasant territories—in order to recognize accurately trade-offs that favorably mediate rural livelihoods, together with the environmental preservation–restoration of the Neotropical landscapes decreed as biosphere reserves.

Elaborated by Tlacaelel Rivera-Núñez

In 2016, a protocol for a 20-hour workshop was designed (Cruz-Morales and García-Barrios 2018) as a means for establishing a dialog with and within an RDG regarding its people, structure, family relations, and lifestyle (values, purposes, and activities); recognizing the strongest external restrictions at the time; and its dynamic strategies for reproducing its capacities year after year. This workshop was carried out twice within that same year (2016). The first workshop was with an RDG from the lower elevations that specializes in dairy farming for producing cheese that is sold regionally (Meza-Jiménez 2016a). The second was with an RDG from the higher altitudes that specializes in producing coffee (Meza-Jiménez 2016b).

In 2017, two workshops were designed and implemented, both of which placed particular emphasis upon inter-generational dialog about the future of social reproduction of RDGs within the CART:

1. A 30-hour workshop entitled "Our mothers and fathers are also our teachers" (Cruz-Morales et al. 2014b). Its purpose was to establish a dialog between children (ages 8–12) and men and women who have distinguished themselves through their interest in maintaining agriculture through stewardship and innovation, as an important resource in the social reproduction of their RDG.

2. A 20-hour workshop entitled "The future of life in the countryside: dialog among women, youths, and men" (Castro-Salcido 2017).

At the end of 2017, six, ten-hour workshops were designed, and then given in many *ejidos*. In those workshops, a total of 126 women, men, and youths carried out simulation exercises with the game *Rio de Vida Campesina* (García-Barrios and Cruz-Morales 2017). Forty-two teams consisting of father, mother, and one son or daughter simulated the social reproduction of their RDG under three conditions: (1) the three members of the team made decisions yearly regarding how to use their capabilities (work, money, land, cities, food supply, and happiness) to achieve social reproduction during at least 10 years (see Cruz-Morales and García-Barrios 2018, for details of the game); (2) the team repeats the previous, but with the option of exchanging tokens (representing capacities) with other teams; and (3) each team proposes and applies three changes to the rules of the game, in order to make the process of reproduction more favorable. Each of the three conditions was accompanied by a collective reflection. An article reporting the results of these workshops is in preparation.

This set of workshops by the AGFAM project was not oriented toward agricultural activities and products, nor was it oriented toward interests and problems of isolated members of the family. Instead, the purpose of AGFAM's workshops was to give women, men, youths, and children an opportunity to reflect together, at a certain depth, upon their values, purposes, agricultural and nonagricultural activities, lifestyles, and interactions, as well their aspirations, hopes, worries, and fears that they experience in their RDGs' daily processes of social reproduction.

Among other things, these family-participation workshops made possible a deeper awareness and understanding of rural families' own perspectives and descriptions of their lives and possibilities, and of their relationship with both agricultural production and the conservation of their agricultural and natural heritage. Our synthesis of some of the relevant facts that we heard from the people themselves now follows.

The adults aspire, repeatedly, to secure a scholarly education for their children that will enable them to be socially mobile. At the same time, the adults voiced their worries not only about the costs and risks of sending their children outside the watershed, but also about the unemployment and underemployment their offspring might encounter after finishing their studies. The majority of RDGs have an agricultural activity that is considered central, but are pluri-active and value the possibility of meeting their own needs for basic grains. The most specialized RDGs prefer to purchase the corn they need from local sources. These RDGs consider that they have average quantities of land and biotic resources, individual and social skills, and physical infrastructure, but that their real bottleneck is the very low availability of money (linked to unfavorable terms of exchange and the lack of credit). A secondary impediment is the relative scarcity of workforce among youth for field labor. The RDGs are averse to borrowing money at usurious rates, and have little capacity for saving. The men are focused intently upon activities that bring monetary income. The majority of women focus on attending to food, clothing, health, and primary education, etc. of members of their RDGs. Youths (irrespective of gender) are less focused, showing diverse and sometimes contrasting interests in three directions: becoming urban professionals, working for wages in northern Mexico or the US, and being agricultural producers in the CART. In the majority of the RDGs, all members participate frequently in religious rituals (now predominantly evangelical), and the adults attend meetings convened by assistance programs.

The participants in family-reflection workshops value resources that are strategic to them, such as water, soil fertility, and availability of firewood, but are quite uninvolved in the project of constructing a REBISE territory, except insofar as that project imposes restrictions upon them. The families see production of shade coffee, resin, palm, and cycads as sources of income, but not as processes of biological conservation. The ways in which the RDGs view their interactions with the numerous external actors vary, and depend upon which actor is involved. In general, the RDGs perceive relations with political parties as more vertical, and less reliable. In contrast, relations with actors who are genuinely interested in conservation and the RDGs' lifestyles are seen as less vertical.

5.4 Conclusions

In this chapter, we have studied conditions and contemporary characteristics of the social reproduction of rural domestic groups (RDGs) in a territory that was declared a protected area in 1995, and that takes in 12 *ejidos* of the *Cuenca Alta del Río el Tablón* (CART) in the Sierra Madre of Chiapas. The RDGs of the territory have had to construct and reconstruct their strategies for social reproduction repeatedly in response to the changing interactions with multiple actors that dispute, in many venues, the territories that RDGs are constructing and the benefits they derive therefrom. In this neoliberal and anti-*campesino* regime, the opportunities and capacities of the territory's RDGs vary markedly. As a consequence, the conditions and social positions of the RDGs are quite diverse, and so (in consequence) are the ways in which they practice agriculture, and in which agriculture contributes to the RDGs social reproduction.

We showed that the first phase in the construction of territory by *campesinos* in the CART (1960–1995) transformed the region profoundly, along with the social condition of its inhabitants—from landless wage laborers to *campesinos*. This first phase occurred before the impacts of neoliberal structural reforms manifested themselves strongly. We showed how those reforms brought rapid, dramatic changes, to which the CART's *campesinos* responded by constructing means of making their continued social reproduction possible under the new conditions. Those means included migration (including remittances), expansion of cattle-raising, and coffee produced in agroforestry for exportation.

We then briefly described and analyzed the actions carried out by two groups of actors: those interested in transforming the territory into a biosphere reserve, and those accompanying the RDGs in responding to the challenges to social reproduction during that period.

We concluded by detailing some of the activities and methods that were designed during 2015–2018 with the express purpose of working with entire RDGs rather than groups consisting of a specific demography (e.g., men, women, and youth). That work was intended to foster dialog among family members by means of very dynamic activities developed during workshops. In the workshops, members discussed their values, hopes, methods, activities, all of which derive from the members' relations among themselves and with other actors. More than 10 years of work in the CART have made possible the creation of trust and accumulation of knowledge (of diverse types) needed for discussing productively with distinct types of RDGs about the challenges they face in social reproduction. We encourage an enduring dialog among the RDGs regarding the changes they must make—with the support of allied actors—to make substantial improvements to their social condition as farmers and as a rural population.

References

Bourdieu P (1990) The logic of practice. Stanford University Press, Stanford

Braasch M, García-Barrios L, Ramírez-Marcial N, Huber-Sannwald E, Cortina-Villar S (2017) Can cattle grazing substitute fire for maintaining appreciated pine savannas at the frontier of a montane forest biosphere-reserve? Agric Ecosyst Environ 250:59–71. https://doi.org/10.1016/j.agee.2017.08.033

Braasch M, García-Barrios L, Cortina Villar S, Huber-Sannwald E, Ramírez-Marcial N (2018) True Grasp: actors visualize and explore hidden limitations of an apparent win-win land management strategy in a MAB reserve. Environ Model Softw 104:153–170. https://doi.org/10.1016/j.envsoft.2018.03.022

Brunel-Manse MC, García-Barrios L (2011) Acknowledging consensus and dissent among and within stakeholder groups over conservation, production and urbanization in a Mexican "Man & the Biosphere" Reserve. Curr Res J Biol Sci 6:457–469

Buhmann K, García-Barrios, L (2014) Aboveground dry matter production and use of fodder tree leaves (Glyricidia sepium), broad-leave herbs and invasive grasses in an experimental on-farm plantation at La Sepultura Bioreserve, Chiapas, Mexico. Bachelor's Dissertation, ECOSUR-Hogeschool VHL University of Applied Sciences

Castro-Salcido E (2017) El futuro del campo: diálogos entre mujeres, hombres y jóvenes. El caso de la cuenca alta del río El Tablón, Sierra de Villaflores, Chiapas. In: Castro-Salcido E (ed) El Colegio de La Frontera Sur Internal Proceedings. San Cristóbal de las Casas

Chayanov AV (1966) In: Thorner D, Smith R, Kerblay B. The theory of peasant economy. Irwin Press, Homewood

Cruz-Morales, J (2014a) Construcción de territorios ambientales mediante procesos de aprendizaje social. El caso de la cuenca alta del río El Tablón, Reserva de la Biosfera La Sepultura, Chiapas, México. PhD Dissertation. Universidad Autónoma Metropolitana, México

Cruz-Morales J (2014b) Desafíos para construir la democracia ambiental en la Cuenca Alta del Río El Tablón (CART), Reserva de la Biosfera La Sepultura (REBISE), Chiapas, México. In: Legorreta-Díaz M, Márquez-Rosano M, Trench T (eds) Paradojas de las tierras protegidas en Chiapas. Democracia y política ambiental en Reservas de Biosfera en Chiapas. Centro de Investigaciones Interdisciplinarias en Ciencias y Humanidades-Centro Regional de Investigaciones Multidisciplinarias-Universidad Nacional Autónoma de México; Dirección de Centros Regionales Universitarios-Universidad Autónoma de Chapingo, México

Cruz-Morales J, García-Barrios LE (2017) Reservas de la Biosfera en Chiapas, México: análisis de las instituciones sociales locales para la conservación y el desarrollo, ¿exclusión y clientelismo? En A. García-García (coord) Extractivismo y neoextractivismo en el sur de México: múltiples miradas. Universidad Autónoma de Chapingo, México

Cruz-Morales J, García-Barrios L (2018) Familia y vida campesina en la Frontera Sur de México: caminos de escucha transdisciplinarios. In: Cruz-Morales J (ed) Familia y vida campesina en la Frontera Sur de México: caminos de escucha transdisciplinarios. El Colegio de la Frontera Sur, Universidad Autónoma Chapingo, San Cristóbal de Las Casas

Cruz-Morales J, Trujillo-Vázquez R, García-Barrios L, Ruiz-Rodríguez JM, Jiménez-Trujillo JA (2011) Buenas prácticas para la ganadería sustentable en la Reserva de la Biosfera La Sepultura (REBISE). Universidad Autónoma Chapingo, El Colegio de la Frontera Sur, Conservación Internacional-México, Comisión de Áreas Naturales Protegidas, San Cristóbal de las Casas

Cruz-Morales J, Fajardo-Piña D, Camacho-Bernal T (2015) Gestión del agua para la ganadería de montaña en la cuenca. In: Rivera Espinosa R (coord) Alternativas sustentables de participación comunitaria para el cuidado del medio ambiente. Servicios Académicos Intercontinentales, Málaga

Dechnik-Vázquez Y, García-Barrios L, Ramírez-Marcial N, van Noordwijk M, Alayón-Gamboa A (2019) Assessment of browsed plants in a sub-tropical forest frontier by means of fuzzy inference. J Environ Manag 236:163–181. https://doi.org/10.1016/j.jenvman.2019.01.071

Figueroa-Sánchez J (2018) Residuos sólidos en Áreas Naturales Protegidas de Chiapas, México ¿Hay gobernanza ambiental para mantenerlas limpias?: el caso del Ejido Los Ángeles, Reserva de La Biósfera La Sepultura. Masters Dissertation, Universidad Autónoma de Chapingo. San Cristóbal de Las Casas

García-Barrios L, Cruz-Morales J (2017) Rio de Vida Campesina. Board game for rural families to explore their social reproduction strategies (not published)

García-Barrios L, González-Espinosa M (2017) Investigación ecológica participativa como apoyo de procesos de manejo y restauración forestal, agroforestal y silvopastoril en territorios campesinos. Experiencias recientes y retos en la sierra Madre de Chiapas, México. Rev Mex Biodivers 88:129–140. https://doi.org/10.1016/j.rmb.2016.10.022

García-Barrios L, García-Barrios R, Waterman A, Cruz-Morales J (2011) Social dilemmas and individual/group coordination strategies in a complex rural land-use game. Int J Commons 5:364–387

García-Barrios L, Álvarez-Solís D, Brunel-Manse MC, Cruz-Morales J, García-Barrios R, Hernández-Ramírez F et al (2012) Innovación socioambiental en la cuenca alta del río El Tablón (CART), sierra de Villaflores, Chiapas. Objetivo, estrategia y métodos de investigación-acción participativa. In: Bello-Baltazar E, Naranjo-Piñera E, Vandame EJ (eds) La otra innovación para el ambiente y la sociedad en la Frontera Sur de México. El Colegio de la Frontera Sur, San Cristóbal de Las Casas, México

García-Barrios L, García-Barrios R, Cruz-Morales J, Smith JA (2015) When death approaches: reverting or exploiting emergent inequity in a complex land-use table- board game. Ecol Soc 20:13–30. https://doi.org/10.5751/ES-07372-200213

García-Barrios L, Cruz-Morales J, Vandermeer J, Perfecto I (2017) The Azteca Chess Experience: learning how to share concepts of ecological complexity with small coffee farmers. Ecol Soc 22:37–57. https://doi.org/10.5751/ES-09184-220237

Gutiérrez-Navarro A (2015) Perspectivas campesinas sobre el uso y manejo del fuego en la Reserva de la Biosfera de la Sepultura, Chiapas. Masters Dissertation, El Colegio de la Frontera Sur, San Cristóbal de Las Casas

IMSS (2014) Instituto Mexicano del Seguro Social, Censo 2014. http://www.imss.gob.mx/. Accessed 30 Nov 2018

Márquez-Conrado R, Cruz-Morales J (2004) Diagnóstico Social y Propuesta de Desarrollo Sustentable para la Reserva de la Biosfera la Sepultura, Chiapas. In: Arellano JL, Monterrosas A, López-Martínez J (coords) 3er Seminario sobre Manejo y Conservación del Suelo y Agua en Chiapas. Instituto Nacional de Investigaciones Forestales, Agrícolas y Pecuarias, The Nature Conservancy, Secretaría de Medio Ambiente y Recursos Naturales, México

Meza-Jiménez A (2016a) Memoria del Primer Encuentro de Familias Ganaderas Lecheras de la Sierra de VIllaflores, Chiapas: nuestra familia, nuestra vida, nuestro futuro. El Colegio de La Frontera Sur Internal Proceedings. San Cristóbal de las Casas

Meza-Jiménez A (2016b) Memoria del Primer Encuentro de Familias Cafetaleras de la Sierra de VIllaflores, Chiapas: nuestra familia, nuestra vida, nuestro futuro. El Colegio de La Frontera Sur Internal Proceedings. San Cristóbal de las Casas

Meza-Jiménez A, García-Barrios L (2015) Los juegos del PAN. Nuevas prácticas docentes para identificar las actitudes, motivaciones y decisiones de la juventud rural a través de juegos del PAN (Patrimonio Agrícola/Natural). Manual del Docente de la Sierra de Villaflores, Chiapas. El Colegio de la Frontera Sur. San Cristóbal de Las Casas, México

Meza-Jiménez A, García-Barrios L, Saldívar-Moreno A, Vera-Noriega JA (2016) Design and evaluation of educational socio-environmental games to identify attitudes, motivations and decisions of smallholder contemporary rural youth. Educare Elect J 20:191–229. https://doi.org/10.15359/ree.20–2.11

Morales-Díaz C (2011) Evaluación experimental de prácticas para establecer plantaciones de Gliricidia sepium en la REBISE, Chiapas. Masters Dissertation, El Colegio de la Frontera Sur, San Cristóbal de Las Casas

Oleta-Barrios J (2012) Adopción del vermicomposteo para establecer árboles forrajeros en la zona de amortiguamiento de la REBISE, Chiapas. México. Masters Dissertation, El Colegio de la Frontera Sur, San Cristóbal de Las Casas

Rosabal-Ayan L (2015) Estrategias campesinas de alimentación del ganado durante el estiaje, en la CART-REBISE, Chiapas, México. Masters Dissertation, El Colegio de la Frontera Sur, San Cristóbal de Las Casas

Sanfiorenzo-Barnhard C (2012) Tree composition and vegetation structure in functional rangelands in a buffer zone of La Sepultura Bio-Reserve Chiapas, Mexico. Masters Dissertation, Universidad de Puerto Rico, Río Piedras, Puerto Rico

Sanfiorenzo-Barnhard C, García-Barrios L, Meléndez-Ackerman E, Trujillo-Vásquez R (2009) Woody cover and local farmers' perceptions of active pasturelands in La Sepultura Biosphere Reserve. Mt Res Dev 29:320–327. https://doi.org/10.1659/mrd.00013

Speelman E, Groot J, García-Barrios L, Kok K, van Keulen H, Tittonell P (2014a) From coping to adaptation to economic and institutional change: trajectories of change in land use management and social organization in a Biosphere Reserve community, Mexico. Land Use Policy 41:31–44. https://doi.org/10.1016/j.landusepol.2014.04.014

Speelman E, García-Barrios L, Groot J, Tittonell P (2014b) Gaming for smallholder participation in the design of more sustainable agricultural landscapes. Agr Syst 126:62–75. https://doi.org/10.1016/j.agsy.2013.09.002

Trujillo-Díaz G, Cruz-Morales J, García Barrios L, Pat Fernández L (2018) Campesinos sin resolución agraria: La difícil construcción de la gobernanza ambiental en una Área Natural Protegida. Revista Pueblos y Frontera digital 13. https://doi.org/10.22201/cimsur.18704115e.2018.v13.335

Valdivieso-Pérez A, García-Barrios L, Plascencia H (2009) Cambio de uso del suelo en la zona de amortiguamiento de la REBISE (1975-2005): crisis del maíz, ganaderización y recuperación arbórea marginal. In: Cavallotti B, Marcof C, Ramírez B (eds) Ganadería y seguridad alimentaria en tiempo de crisis. Universidad Autónoma Chapingo, México

Valdivieso-Pérez A, García-Barrios L, Álvarez-Solís D, Nahed-Toral J (2012) De maizales a potreros: cambio en la calidad del suelo en Los Ángeles, Villaflores, Chiapas, México. Terra Latinoamericana 30:363–374

Valencia V, García-Barrios L, West P, Sterling E, Naeem S (2014) The role of coffee agroforestry in the conservation of tree diversity and community composition of native forests in a Biosphere Reserve. Agric Ecosyst Environ 189:154–163. https://doi.org/10.1016/j.agee.2014.03.024

Valencia V, West P, Sterling EJ, García-Barrios L, Naeem S (2015) The use of farmers' knowledge in coffee agroforestry management: implications for the conservation of tree biodiversity. Ecosphere 6:1–17. https://doi.org/10.1890/ES14-00428.1

Valencia V, Naeem S, García-Barrios L, West P, Sterling EJ (2016) Conservation of tree species of late succession and conservation concern in coffee agroforestry systems. Agric Ecosyst Environ 219:32–41. https://doi.org/10.1016/j.agee.2015.12.004

Valencia V, García-Barrios L, Sterling EJ, West P, Meza-Jiménez A, Naeem S (2018) Smallholder response to environmental change: impacts of coffee leaf rust in a forest frontier in Mexico. Land Use Policy 79:463–474. https://doi.org/10.1016/j.landusepol.2018.08.020

Vides-Borrell E, García-Barrios L, Álvarez-Solís D, Nigh R, Astier-Calderón M, Douterlungne D (2011) Survival and early growth of Gliricidia sepium fodder trees in subhumid tropical pasturelands: contrasting effects of NPK fertilizer salts vs. organic ammendments. Res J Biol Sci 6:468–474

Zabala A (2015) Rural livelihoods and attitudes toward silvopastoral innovations in a buffer zone of La Sepultura Bio-Reserve, Chiapas, Mexico. PhD Dissertation, Cambridge University, United Kingdom

Zabala A, Pascual U, García-Barrios L (2017) Payments for pioneers? Revisiting the role of external rewards for sustainable innovation under heterogeneous motivations. Ecol Econ 135:234–245. https://doi.org/10.1016/j.ecolecon.2017.01.011

Chapter 6
Macroalgae Mariculture as a Social and Environmental Alternative for Coastal Traditional Communities of the Semi-Arid Region of Northeast Brazil

Ivanilson de Souza Maia, Dárlio Inácio Alves Teixeira, Vigínia Maria Cavalari Henriques, and Maulori Curié Cabral

6.1 Introduction

Artisanal fishing along the semi-arid coast of Northeast Brazil suffers the same consequences as the rest of the world: diminishing yields, reduced stocks, fragile returns, impoverished fishing communities, and a sense of abandonment (Souza Maia and Oliveira-Neto 2012).

Despite this situation, fishing is the main source of protein and income in the Northeast Region, with just over 40% of all Brazilian fishers (BRASIL 2006). Among the fishing segments in coastal areas is macroalgae collecting, which, in the Northeast, is marked by the presence of females.

The theme of gender and fishing has been little explored in the academic world, which is an intriguing problem because fishing is still considered a male practice in spite of the participation of women in different activities of the production chain (Leitão 2009).

Traditional communities have long depended on natural banks as their main source for acquiring macroalgae. Intensive and disorderly collection of macroalgae,

I. de Souza Maia (✉)
Departamento de Ciência Animal, Centro de Ciências Agrárias, Universidade Federal Rural do Semi-Árido (UFERSA), Mossoró, RN, Brazil
e-mail: ivanilson.maia@ufersa.edu.br

D. I. A. Teixeira
Escola Agrícola de Jundiaí/Unidade Acadêmica Especializada em Ciências Agrárias, Universidade Federal do Rio Grande do Norte (EAJ/UFRN), Macaíba, RN, Brazil

V. M. C. Henriques
Departamento de Oceanografia e Limnologia, Centro de Biociências, Universidade Federal do Rio Grande do Norte (UFRN), Natal, RN, Brazil

M. C. Cabral
Instituto de Microbiologia Paulo de Góes, Centro de Ciências da Saúde, Universidade Federal do Rio de Janeiro (UFRJ), Rio de Janeiro, RJ, Brazil

C. Baldauf (ed.), *Participatory Biodiversity Conservation*,
https://doi.org/10.1007/978-3-030-41686-7_6

especially species with economic potential, has led to considerable declines of native species in the marine environment (Souza Maia et al. 2011).

Marine macroalgae are benthic photosynthetic organisms that are involved in nutrient cycling as one of the bases of the trophic chain. They are of extreme environmental importance because they prevent marine erosion while providing shelter for various aquatic species (Câmara-Neto 1971).

Macroalgae mariculture is an alternative that can reduce the intensity of algal collection from natural banks and avoid exploitation beyond carrying capacity. It also offers a number of advantages such as easy deployment and harvesting, product control, higher productivity and specificity, and production adjustment to market demand (FAO 2016). Furthermore, macroalgae mariculture requires little investment and can even be developed in polyculture with shrimp (Alencar et al. 2010). The most cultivated species in the region is *Gracilaria birdiae*.

The present work highlights several possibilities for the use of marine macroalgae, from dry or powdery preparation to the use of hydrolyzate, which, as marine algae juice, can reduce or replace the abundant and widespread use of agricultural synthetics. The rational use of this juice is an economic opportunity for family mariculture and of relevance to aspects of the environment.

This work aimed to present the potential that marine macroalgae mariculture holds for the coast of the semi-arid region of Northeast Brazil, with regard to local development and the elaboration of by-products of economic importance.

6.2 Characteristics of the Study Area

The study was conducted in the southern semi-arid portion of the Northeast coast of Brazil, wedged between Cabo Calcanhar, in Touros, state of Rio Grande the Norte (RN) at 35°27′W, and Ponta de Itapajé, state of Ceará (CE) at 40°W (Muehe and Garcez 2005). According to Duque (1953), this portion of the semi-arid Northeast is unique as it embraces the sea.

Climatic factors, associated with geomorphological and phytogeographic attributes, characterize the area, mainly due to rainfall being below evapotranspiration. The climate of the region is classified by Köppen (1948) as semi-arid type BShs'w'—hot tropical bioclimate with accentuated drought. The rainy season occurs between January and July, with the highest rainfall being concentrated in March and April (Silva et al. 2011), being associated with Zona de Convergência Intertropical (ZCIT: Intertropical Convergence Zone) (Testa and Bosence 1999; Jimenez et al. 1999). The mean annual temperature is 28 °C, with the dynamics of air masses influencing the emergence and regime of winds and ocean circulation patterns, which shape the coast (Vital 2006).

This area encompasses 584 km of coastline and occupies more than 17% of the Northeast coast (Pinheiro et al. 2008). Among the communities that practice macroalgae mariculture in this area, Rio do Fogo, RN, and Barrinha/Icapuí, and Flecheiras/Trairí, CE, in particular, have collaborated in its development.

The three areas under study present satisfactory conditions for the development of macroalgae mariculture since they are physically protected from currents, winds, and waves, and are near seaweed banks, which guarantee seed stock for cultivation.

According to Muehe and Garcez (2005), low productivity predominates the continental shelf in the area due to reduced nutrient availability, chlorophyll "*a*" concentration below 0.1 mg/m^3 and a narrow continental margin.

Importantly, the availability of nutrients influences not only the low productivity but also the type of sedimentary cover and the width of the continental margin itself. Souza Maia and Oliveira-Neto (2012) reported that the reduced availability of nutrients along the coast of the semi-arid Northeast is due to the fact that the rivers originate in the semi-arid region and possess numerous dams, which decisively contribute to negative discharges into the sea.

6.3 Environments, Territorialities, and Resilience

The potential offered by coastal environments has always aroused human interest in the use and occupation of land adjacent to the sea. According to Strohaecker (2008), this occurred, in principle, for the purpose of survival and sustenance of tribal communities, but over time for the conquest of new territories with the aim of obtaining wealth and power.

Marine and coastal environments in this area, as well as in Brazil, promote opportunities for economic and social activities due to their characteristics and attributes (Souza Maia and Oliveira-Neto 2012). These environments are mainly used for urbanization, tourism, agriculture and agribusiness, extractivism, petroleum activities, renewable energy, ports, fishing, and aquaculture (Vasconcellos et al. 2007; Anello 2009; Vasconcellos 2012; Knox and Trigueiro 2015). Economic activities developed on the Brazilian coast are responsible for about 70% of national GDP (Serafim and Hazin 2005). This potential has transformed the Brazilian coastal zone into a scene of various forms of occupation and use, conflicts, and development.

According to Silva (2003) and Cabral et al. (2015), the creation of a new coastal/seaside world—complex, conflicting and changing—is no different from other territories impacted by globalization but is a major challenge for sustained management that benefits the lives of traditional residents and their families.

Fishing, including family mariculture, one of the most important sources of income of coastal populations, has resisted the transformations caused by the dynamics of capitalist activities, given the state of exploitation of natural resources (Diegues 1995).

Marine macroalgae cultivation avoids harvesting from natural banks, the biomass of which has high carbon sequestration power. It also creates an environment unconducive to breeding and sheltering of the of lobster puerulus, shrimp larvae, juvenile seahorses, and several species of fish, which is an important environmental service for artisanal fishing (Fig. 6.1).

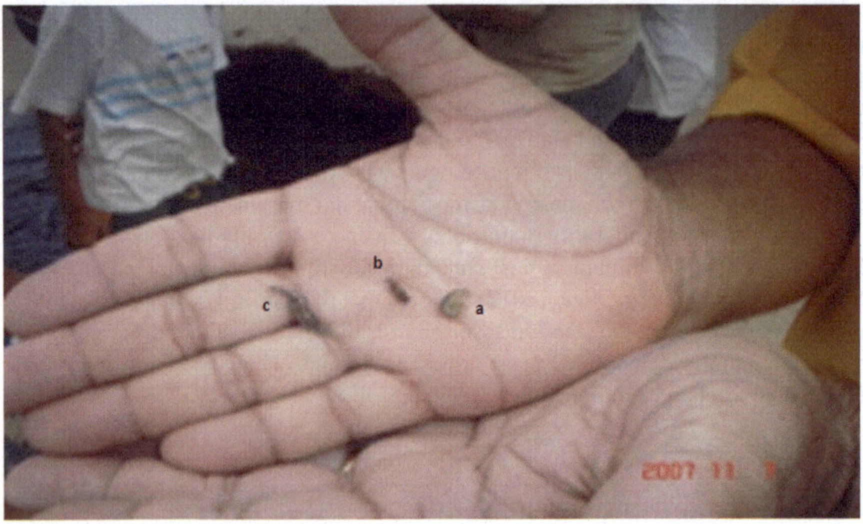

Fig. 6.1 Lobster puerulus (**a**), shrimp larvae (**b**) and juvenile seahorse (**c**) found in macroalgae cultivation. (Photo: Ivanilson de Souza Maia)

Artisanal fishing and family mariculture strengthen resilience by acting as a safety net during times of scarcity and catastrophe when other food production sectors do not function (FAO 2018). Resilience, understood from the concept of Holling (1973), is as follows:

> Emergent property of systems and conceived as the intensity and frequency of disturbances that a socio-ecological system can absorb without undergoing fundamental changes in its functional characteristics.

Family marine macroalgae mariculture has shown a pattern in the strategies it has adopted as a way of interacting with the marine system and, at the same time, adaption to systematic changes occurring at different scales in this ecosystem. These strategies were relevant to strengthening local management.

Thus, it is clear that family marine macroalgae maricutlurists develop a set of strategies that promote resilience, from the socio-ecological point of view of production, such as traditional knowledge of work, shared capital, social and environmental responsibility, and leadership. These factors constitute spaces for learning about the socio-ecological processes that have been fundamental for the institutional growth of all organizations of mariculturists and their empowerment in the community.

6.4 Identity, Female Empowerment, and Local Development

Women who were once collectors of "cisco"—terminology used in traditional coastal communities for seaweed—using techniques passed on by grandparents, mothers or friends, have now become family marine macroalgae maricutlurists.

Strong identity with the activity, the family relationship with colleagues and a series of interventions, guidance and follow-ups by universities and third sector organizations with projects to cultivate marine macroalgae, attracted women to cultivation. In the beginning, they faced difficulties with cultivation technology due to the environmental conditions of the place and the state change from extractivist to cultivator. In the course of cultivation, and with the help of technical support, they were able to develop a technology that suited local conditions.

Their need to earn an income contributed significantly to their cultivation of marine macroalgae. Another important factor was the social and environmental condition they faced in their communities. They saw in marine macroalgae mariculture the baton for their organization, the empowerment of women, and community leadership.

Internal conflicts occur when there is non-compliance with their established rules of coexistence, especially with regard to non-compliance with established times for cultivation and processing. Externally, they occurred primarily in the early years by family members, neighbors, and politicians.

Regarding politicians, however, it still occurs less frequently. It was evident in a veiled dispute for power between the city council and the association of women mariculturists. Mainly, referring to the donation of land to the association, the receipt of external institutional visits or the invitation of these institutions to the mariculturists to represent the municipality in a national event.

They also faced other conflicts of environmental nature with fishermen who poured engine and gearbox oil of the boats into the sea. The oil spills moved toward the crop. It took several assaults on them to gain their respect. Observations of a mariculturist: "At first we would fight, then we would convince them. It shows how polite we are."

Participation was the driver for awakening the citizenship employed in the situations expressed above. They are members of the Territory Council of Fisheries and Aquaculture Citizenship, Environmental Protection Area, and the Municipal Council of Child and Adolescent.

The motive for participation was exchange at the state and national level, where they participated by orally presenting their projects in seminars on macroalgae mariculture in Brasília, Santa Catarina, São Paulo, Fortaleza, Natal, Recife and Salvador, as well as at Conferência de Pesca e Aquicultura (Fisheries and Aquaculture Conference) at state and national levels. This, in turn, transformed into pride for local society in the municipalities.

The set of institutions made it possible to complement project advances, such as environmental licensing and the granting of area for 20 years to explore cultivation, tax rate reduction, research, and technological development.

Of course, all of this was only possible because the groups had organizational discipline and a willingness to win. They had regular monthly meetings, and met other times as needed. They have a fund with individual contributions and a record of minutes for all meetings.

Marine macroalgae mariculturists currently sell algae as dehydrated branches or in crushed (powered) form. During the processing of these products, 100% of the liquid nutrients of alae are wasted. This hydrolyzate represents a product of high commercial value when used as feedstock for food production, with biosecurity and significant nutritional value for livestock, especially monogastric animals.

Technological development and innovation can create opportunities for these communities, enabling them to achieve local development and sustainability of family marine macroalgae mariculture, while contributing decisively to artisanal fishing.

6.5 Technologies and Perspectives for Communities

New equipment for cultivation includes floating rafts with tubular nets, which replaced the ropes used to fix seedlings. This change was well accepted by project participants since it was an easy-to-use technique with low aggregate technology and provided a higher speed of field work and reduced losses of crop seedlings, which had been detrimental to offshore cultivation.

Marine macroalgae cultivation contributes to the reproduction of species and provides shelter and protection for young forms. It also provides structure that mitigates environmental impacts. A 100 ha *Gracilaria* (marine macroalgae) farm can, theoretically, remove 80% of the dissolved nitrogen released by a 1500-ton salmon farm (Buschmann et al. 2008; Abreu et al. 2009).

Research carried out in partnerships between universities and mariculturist associations has shown beneficial effects for agriculture from the use of marine algae extracts. Likewise, Button and Noyes (1964) observed increased seed germination and plant growth.

6.6 Final Considerations

Given the current situation of fishing communities, it is necessary and urgent to implement experimental crops that can serve as practical subsidies for the development and implementation of family cultivations, as has been done in some Asian countries.

Macroalgae cultivation could also serve as an excellent rehabilitation program for poor coastal communities. This type of action can become an option to alleviate the poverty of fishers, while contributing to the restoration of natural banks.

Considering these relevant facts regarding technology, low primary production and thinking of generating a decent income for mariculturists, it is possible to apply cultivation in the form of periodic work, alternating cultivation with responsible sustainable management of natural banks within current legislation.

References

Abreu ML, Varela DA, Henríquez L, Villarroel A, Yarish C, Sousa-Pinto I, Buschmann AH (2009) Traditional vs. integrated multi-trophic aquaculture of Gracilaria chilensis. C. J. Bird, J. McLachlan & E.C. Oliveira: productivity and physiological performance. Aquaculture 293:211–220. https://doi.org/10.1016/j.aquaculture.2009.03.043

Alencar R, Horta PA Jr, Celino JJ (2010) Cultivo de camarão branco *Litopenaeus vannamei* (Boone, 1931) com a macroalga *Ulva Lacuata* Linneaus (Chlorophyta) no tratamento de efluentes em sistema fechado de recirculação. Revista de Biologia e Ciências da Terra 10(1):117–137

Anello LFS (2009) Os programas de educação ambiental no contexto das medidas compensatórias e mitigadoras no licenciamento ambiental de empreendimentos de exploração de petróleo e gás no mar do Brasil: a totalidade e a práxis como princípio e diretriz de execução. Dissertation, Universidade Federal do Rio Grande

BRASIL (2006) Resultados do recadastramento nacional dos pescadores do Brasil. Brasília, Distrito Federal, Secretaria da Administração e Previdência, Presidência da República

Buschmann AH, Varela DA, Hernández-González MC, Huovinen P (2008) Opportunities and challenges for the development of an integrated seaweed-based aquaculture activity in Chile: determining the physiological capabilities of *Macrocystis* and *Gracilaria* as biofilters. J Appl Phycol 20:571–577

Button EF, Noyes CF (1964) Effect of seaweed extract upon emergence and survival of seedlings of creeping red fescus. Agron J 56:444–445

Cabral NWSS, Coelho GC, Soares MLS (2015) Zona Costeira resiliente: um estudo socio-ecológico no Nordeste paranse. Anais do IV International Symposium on Project Management, Innovation and Sustainability, São Paulo, ISSN: 2317–8302

Câmara-Neto C (1971) Contribuição ao conhecimento qualitativo e quantitativo das "arribadas" da Redinha. Boletim do Instituto de Biologia Marinha 5:3–30

Diegues ACS (1995) Povos e mares: leitura em sócio antropologia marítima. Núcleo de Apoio à Pesquisa sobre Populações Humanas em Áreas Úmidas Brasileiras, Universidade de São Paulo, São Paulo

Duque JG (1953) Solo e a água no polígono das secas. Departamento Nacional de Obras Contra as Secas, Fortaleza

FAO (2016) The State of World Fisheries and Aquaculture 2016 (SOFIA): contributing to food security and nutrition for all. Food and Agriculture Organization, Rome

FAO (2018) El estado mundial de la pesca y la acuicultura: cumplir los objetivos de desarrollo sostenible. Organización de las Naciones Unidas para la Alimentación y la Agricultura, Roma

Holling CS (1973) Resilience and stability of ecological systems. Annu Rev Ecol Syst 4:1–23

Jimenez JA, Maia LP, Serra J, Morais J (1999) Aeolian dune migration along the Ceará coast, north-eastern Brazil. Sedimentology 46(4):689–701

Knox W, Trigueiro A (2015) A pesca artesanal do litoral do ES. In: Knox W, Triguerio A (orgs) Narrativas saberes e conflitos na pesca artesanal. Editora da Universidade Federal do Espírito Santo, Vitória

Köppen W (1948) Climatología: com un estudio de los climas de la tierra. Fondo de Cultura Econômica, México, 478 p

Leitão MRFA (2009) Gênero e políticas públicas na pesca artesanal de Itapissuma. In: Callou ABF, Tauk Santos MS, Gehlen VRF (orgs) Comunicação, gênero e cultura em comunidades pesqueiras contemporâneas. Fundação Antônio Santos Abranches, Recife

Muehe D, Garcez DS (2005) A plataforma continental brasileira e sua relação com a zona costeira e a pesca. Mercator 4(8):69–88

Pinheiro LS, Coriolano LN, Costa MF, Dias JA (2008) O Nordeste brasileiro e a Gestão Costeira Integrada. Revista de Gestão Costeira 8(2):5–10. https://doi.org/10.5894/rgci58

Serafim CFS, Hazin F (2005) O ecossistema costeiro. In: Chaves PT (org) Geografia: ensino fundamental e médio: o mar no espaço geográfico brasileiro. MEC, Brasília, pp 101–131

Silva J (2003) A extensão Pesqueira" no desenvolvimento das comunidades litorâneas no Estado do Ceará. Thesis, Universidade Federal do Ceará, Fortaleza

Silva VPR, Pereira ERR, Azevedo PV, Souza FAS, Sousa IF (2011) Análise da pluviometria e dias chuvosos na região Nordeste do Brasil. Revista Brasileira de Engenharia Agrícola e Ambiental 15(2):131–138

Souza Maia I, Oliveira-Neto JT (2012) Estudo de viabilidade econômica e gestão democrática de empreendimentos populares: o caso das marisqueiras do semiárido potiguar. Vivência: Revista de Antropologia 40:67–79

Souza Maia I, Bezerra AF, Cavalari VMH (2011) Avaliação de sustentabilidade em empreendimentos popular: o caso das mulheres das a(l)g(u)as de Rio do Fogo, RN. III Workshop Interdisciplinar de Pesquisa em Indicadores de Sustentabilidade, São Carlos: EESC/FSP/USP 2:40–42

Strohaecker TM (2008) Dinâmica populacional. In: BRASIL. Ministério do Meio Ambiente. Macrodiagnóstico da Zona Costeira e Marinha. IBAMA/MMA, Brasília, pp 59–73

Testa V, Bosence DWJ (1999) Physical and biological controls on the formation of carbonate and siliciclastic bedforms on the northeast Brazilian shelf. Sedimentology 46(2):279–301

Vasconcellos LG (2012) Pesca artesanal e petróleo no Recôncavo Baiano: gestão ambiental federal como mediadora de conflitos. Revista Nordestina de Ecoturismo 5(1):103–110

Vasconcellos M, Diegues ACSA, Sales RR (2007) Limites e possibilidades na gestão da pesca artesanal costeira. In: Costa AL (ed) Nas Redes da Pesca Artesanal. IBAMA - MMA, Brasília, pp 15–83

Vital H (2006) Erosão e progradação do litoral brasileiro - capítulo: Rio Grande do Norte. In: Muehe D (ed) Erosão e progradação do litoral brasileiro. MMA, Brasília, 476 p

Chapter 7
Community-Based Management of Amazonian Biodiversity Assets

João Vitor Campos-Silva, Joseph E. Hawes, Carolina T. Freitas, Paulo C. M. Andrade, and Carlos A. Peres

7.1 The Amazon Socioecological System

The Amazon basin hosts about half of the remaining tropical forests on Earth (Hansen et al. 2013) and a huge diversity of freshwater and terrestrial biodiversity. In addition, it also sustains a multiplicity of human societies, with an impressive linguistic and cultural diversity that evolved in a complex landscape pervaded by an extensive network of rivers (Arias et al. 2018). In parallel, the Amazon forest also represents a great temptation to the governments of nine South American

J. V. Campos-Silva (✉)
Instituto de Ciências Biológicas e da Saúde, Universidade Federal de Alagoas, Maceió, AL, Brazil

Faculty of Environmental and Natural Resource Management, Norwegian University of Life Sciences, Ås, Norway

J. E. Hawes
Faculty of Environmental and Natural Resource Management, Norwegian University of Life Sciences, Ås, Norway

Applied Ecology Research Group, School of Life Sciences, Anglia Ruskin University, Cambridge, UK

C. T. Freitas
Departamento de Ecologia, Centro de Biociências, Universidade Federal do Rio Grande do Norte, Natal, RN, Brazil

Divisão de Sensoriamento Remoto, Coordenação de Observação da Terra, Instituto Nacional de Pesquisas Espaciais, São José dos Campos, SP, Brazil, Brazil

P. C. M. Andrade
Departamento de Produção Animal e Vegetal, Laboratório de Animais Silvestres, Universidade Federal do Amazonas, Manaus, AM, Brazil

C. A. Peres
School of Environmental Sciences, University of East Anglia, Norwich Research Park, Norwich, UK

C. Baldauf (ed.), *Participatory Biodiversity Conservation*,
https://doi.org/10.1007/978-3-030-41686-7_7

countries in terms of industrial development, due to immeasurable high-value natural resources occurring within their boundaries. Therefore, reconciling sustainable pathways for biodiversity conservation, poverty alleviation, social insurance of traditional communities and economic growth, will determine the fate of the largest tropical forest on Earth.

The socioecology of the Amazon is highly diverse, formed not only by a variety of indigenous groups but also by small farmers, extractivists, and artisanal fishers, each with different cultures and relations with the forest and its natural resources (Lima and Pozzobon 2005). Biodiversity and ecosystem services represent the cornerstone of Amazonian cultural maintenance, which is intertwined with traditional activities such as fishing, hunting, and harvesting of forest products (Sunderlin et al. 2005). In this context, effective management of natural resources is one of the most imperative local demands, but also a challenging task, due to severe shortages of research funding and human resources (Campos-Silva et al. 2015; Magnusson et al. 2018), and often unrealistic expectations about what can be defined as demographically sustainable exploitation (Peres 2011; Terborgh and Peres 2017).

Protected Areas (PAs) represent the main existing strategy to protect natural resources, biodiversity, and traditional livelihoods (Bruner et al. 2001; Coetzee et al. 2014; Watson et al. 2014). However, tropical PA systems in developing countries, such as Brazil, face particular challenges, which strongly compromise their effectiveness against the powerful drivers of deforestation, overexploitation, and habitat destruction from mining, dams and other developments (Bruner et al. 2004). A clear example is the PA system in State of Amazonas (Brazilian Amazon), where only three staff are designated to manage all 42 state PAs, which corresponds to one park manager for every 6.3 Mha of PAs (Campos-Silva et al. 2017). This current PA system is evidently not enough to ensure the conservation of the Brazilian Amazon, and new strategies should be urgently designed.

Formal partnership with local communities, such as community-based management (CBM) or co-management arrangements, represent a promising alternative to increase local governance, decentralize decision making, strengthen surveillance systems, and reduce conservation costs (Somanathan et al. 2009). There are positive examples worldwide, where local communities play a central role in local resource management (e.g., Gibson and Marks 1995; Cinner et al. 2012a, b). CBM has been particularly successful in the Amazon, promoting strong ecological benefits, ensuring the conservation of a large set of taxonomic groups and the population recovery of overexploited species (Castello et al. 2009; Campos-Silva and Peres 2016; Petersen et al. 2016; Campos-Silva et al. 2017, 2018). At the same time, CBM initiatives have strongly contributed to the development of rural communities, improving many mainstream and unconventional socioeconomic indicators, including income generation, community pride, and maintenance of cultural capital (Campos-Silva and Peres 2016).

7.2 CBM of Biodiversity Assets as Bright Spots in the Amazon

Biodiversity assets can be understood as any resource or relationship that can be managed or protected, generating forms of values for individuals, communities, or institutions (Jepson et al. 2017). Beyond the material value, represented by income generation, food security and resources used on a daily basis, the assets concept also incorporates immaterial values that add emotion, pride, sense of justice, and different cosmologies that are not necessarily aligned with a hegemonic neoliberal paradigm. These subjective values are important to strengthen collective actions capable of shaping the social dynamic within broad socioecological systems (Jepson et al. 2017).

The Amazon basin hosts at least two successful CBM initiatives that have led to positive outcomes for both biodiversity conservation and local well-being. These examples can be considered as "bright spots" able to inspire new pathways where biodiversity conservation and local needs are truly aligned. Focusing on these bright spots is a powerful ingredient to build conservation optimism, which can strongly influence decision makers and stakeholders (Cvitanovic and Hobday 2018). These positive examples have high potential for being scaled up throughout Amazonian floodplains and to become an effective and decentralized conservation tool at vast spatial scales. Here, we elucidate how two examples of CBM have been working to date along the Juruá River, a major tributary of the Amazon River. We describe the operational mechanisms of these impressive examples, identifying the main ingredients to consolidate a robust CBM program, which can be applied in different contexts to achieve strong social and ecological outcomes elsewhere.

7.3 Socioecological Context of the Juruá River

The Juruá River is characterized by its large highly productive floodplains, sustaining hundreds of indigenous and nonindigenous human settlements (Fig. 7.1). The landscape is comprised of seasonally flooded (*várzea*) forests across the whole floodplain and adjacent upland (*terra firme*) forests (Hawes et al. 2012). The Juruá, especially the middle section of this river, played an important role during the rubber boom, where thousands of people from northeast Brazil migrated to the Amazon to work as rubber tappers. These people lived under conditions that almost amount to slavery, without social rights, and often suffering from severe destitution, debt patronage, tropical diseases, and lack of access to health and education (Almeida 2002). With the help of the Catholic Church and the environmental movement that took shape around the public figure of the social activist Chico Mendes, these local communities started a process of self-organization to ensure essential social and land rights (Fearnside 1989). In this context, two large sustainable-use PAs were created in this region. The federally managed Médio Juruá Extractive Reserve

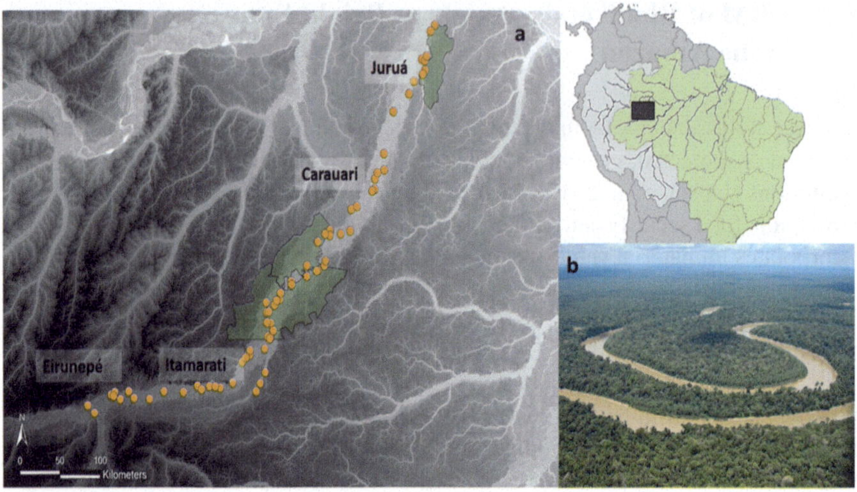

Fig. 7.1 (**a**) Distribution of protected areas across the Juruá River in western Brazilian Amazonia. The orange circles represent the rural communities. Green polygons represent three sustainable-use forest reserves. (**b**) The meanders of Juruá River

(ResEx Médio Juruá; 5°33′54″S, 67°42′47″W) was created in 1997 and hosts some 700 people distributed across 13 villages in its 253,227 hectares. The state-managed Uacari Sustainable Development Reserve (RDS de Uacari; 5°43′58″S, 67°46′53″W) hosts about 1200 villagers living in 32 communities within its 632,949 hectares. The local economy in both reserves is sustained by fisheries, slash-and-burn agriculture, and nontimber forest products, such as oil seeds and palm fruits (Newton et al. 2011), and supported by payments for environmental services (Alves-Pinto et al. 2018), but two examples of CBM stand out for generating broad social and economic benefits for rural communities: the CBM of arapaima and freshwater turtles.

7.4 Community-Based Management of Arapaima

Arapaima (*Arapaima gigas*), also referred to as pirarucu or paiche, is the largest freshwater scaled fish on Earth, reaching up to 3 m in length and over 200 kg in weight (Nelson 1994). This iconic species has played a central subsistence role in the Amazon since pre-Columbian times (Prestes-Carneiro et al. 2016). However, over the last century arapaima populations suffered a dramatic decline due to intensive commercial pressure, and were extirpated in many areas (Castello et al. 2015). Arapaima fishery was then banned by the Brazilian government, yet such measure was not enough to ensure recovery due to widespread high levels of illegal fishing (Castello and Stewart 2010; Cavole et al. 2015). In an attempt to reverse the arapaima population collapse, local communities, experienced fishers, and researchers first started a CBM model in 1999 at the Mamirauá Sustainable Development Reserve (Castello et al. 2009, 2011).

Arapaima evolved in an anoxic lake environment, and natural selection led to changes in the swim bladder, which became adapted for breathing (Brauner 2004). As a result, arapaima regularly comes to the surface to capture oxygen directly from the air, and individuals can therefore be visually counted by trained fishers following a standardized protocol (Castello 2004). This natural feature has broad implications for harvesting management; it allows the population size of arapaima to be reliably estimated, and from these estimates the government assigns an undifferentiated harvest quota of up to 30% of adult individuals in each CBM unit (Castello et al. 2011). Another important feature of the CBM scheme is that local communities need to zone their water bodies, including areas for protection (protected no-take lakes), and to ensure full-time local surveillance against poaching. This initiative proved to be highly successful (Castello et al. 2009), and as a consequence arapaima CBM schemes have since spread to other areas across the Amazon.

Studies have shown that arapaima CBM has been generating impressive outcomes in several sites across the Amazon, for both biodiversity conservation and the well-being of rural communities (Castello et al. 2009; Campos-Silva and Peres 2016; Petersen et al. 2016). Along the Juruá River, in western Brazilian Amazonia, community-based protection of lakes has induced a huge recovery of arapaima populations, with an increase of 425.2% within 11 years of CBM (Campos-Silva et al. 2019). Even outside PAs, arapaima population trends show the same pattern, increasing 397.5% per year (Campos-Silva et al. 2019). A single protected lake can host more than 2800 individuals, while unprotected lakes on average only support nine individuals (Campos-Silva and Peres 2016; Campos-Silva et al. 2019), which underlines the high success of this model. The same pattern was found in other river basins, where arapaima has also been brought back from the brink (Castello et al. 2009; Petersen et al. 2016). In addition to conservation gains for arapaima, community-based protection of lakes also benefits a large set of co-occurring species, including caimans, freshwater turtles, and other fish species (Miorando et al. 2013; Campos-Silva and Peres 2016; Arantes and Freitas 2016).

Furthermore, beyond these ecological outcomes, arapaima CBM has also triggered a substantial social transformation in Amazonia, through improvements in rural well-being. First, protected lakes ensure an annual income to rural people typically short on other options for earning cash. The social security that this provides allows revenues to be saved and used in cases of emergency, for example, urgent health care (Campos-Silva and Peres 2016). Second, profits secured from the harvest can also help improve basic infrastructure and living conditions in households and communities (Campos-Silva and Peres 2016). Other important social outputs perceived by people actively participating in arapaima CBM are improved food security, community pride, cultural maintenance, and a more equitable distribution of profits from fisheries (Campos-Silva and Peres 2016).

7.5 Community-Based Management of Amazonian Freshwater Turtles

Freshwater turtles, particularly those from the genus *Podocnemis*, including *P. expansa, P. unifilis* and *P. sextuberculata*, also show a high cultural value throughout Amazonian history by both indigenous and nonindigenous dwellers (Prestes-Carneiro et al. 2016). Recorded uses other than the consumption of meat and eggs, include fat to produce oil for fuel, medicine (e.g., fat and bone), and the carapace for ritual purposes and as a tool (Smith 1979; Rebêlo and Pezzuti 2000; Pezzuti et al. 2010). As for arapaima, many freshwater turtles and especially the Giant South American turtle (*P. expansa*) experienced huge population declines over a large geographic scale due to historical commercial overexploitation (Schneider et al. 2011).

In an attempt to reverse the imminent population collapse in turtle stocks and to safeguard the high-value resource they potentially provide, the Brazilian government in partnership with local communities started a process of CBM in the early 1970s, focused on the protection of fluvial beaches which *P. expansa* (and other *Podocnemis* species) use to nest (Andrade 2007; Cantarelli et al. 2014). Freshwater turtles are still considered a high-value delicacy among rural and urban people (Rebêlo and Pezzuti 2000), and their conservation strongly depends on local support in about 390 protected beaches across Brazilian Amazonia. At each of these beaches, beach guards ensure local surveillance throughout the breeding season, which results in reduced poaching of nesting females and their eggs (Campos-Silva et al. 2018).

Similarly to arapaima CBM, community-based beach protection has led to strong ecological outcomes. For example, after 40 years of CBM along the Juruá River, populations of freshwater turtles are in a recovery process (Campos-Silva et al. 2018). The number of Giant South American turtles nesting there is 58-fold higher on protected beaches compared to unprotected beaches, and nest poaching is about 2% and 99% in the protected and unprotected beaches, respectively (Campos-Silva et al. 2018). Collateral benefits from beach protection are also widespread across other taxonomic groups, including caimans, birds, iguanas, fishes, and even invertebrates (Campos-Silva et al. 2018), reinforcing the role of freshwater turtles as an umbrella species through beach protection.

An important difference between the CBM of arapaima and freshwater turtles is the current lack of economic returns in the latter. In our study region along the Juruá River, beaches are guarded 24/7 throughout the entire breeding season (5 months), yet beach guards receive only around U$ 110 in the form of a food hamper (Campos-Silva et al. 2018). Considering the high personal life risk undertaken, and the physical and mental effort required for effective protection, it is clear that beach guards are severely underpaid for their services. Despite their dedication to the cause, dissatisfaction among beach guards is increasing, threatening the continuity of this successful program over the long term (Campos-Silva et al. 2018).

7.6 Principles Ensuring Success of Amazonian Bright Spots

The literature to date identifies a set of important principles for achieving the cohesive management of common natural resource pools at the local community level (Ostrom 2009; Cox et al. 2010, 2016). These principles often occur in successful projects that deliver positive outcomes for both biodiversity conservation and local welfare (Pinkerton and Weinstein 1995; Castello et al. 2009; Cox et al. 2010; Gutiérrez et al. 2011; Campos-Silva and Peres 2016). From a literature review of such recognized principles on common pool resource management and our own collective experience in Amazonian CBM programs, we present a model comprising the social and institutional principles, and the intrinsic values inherent in biodiversity assets, which have been used by traditional communities for centuries (Fig. 7.2). This model can help strengthen existing CBM programs in Amazonia and inspire new initiatives.

As the first principle, the presence of leadership and social capital seems to be fundamental in successful examples of CBM (Gutiérrez et al. 2011). Strong leadership, defined as an individual with entrepreneurial behavior and high motivation who is respected as a local leader, can inspire behavioral changes within the wider community, increasing local engagement, commitment, and collective compliance of other residents (Gutiérrez et al. 2011). However, it is very important that local leaders do not use their privileged condition and access to economic opportunities for self-benefits, which can erode their own legitimacy (Muehlig-Hofmann 2007). Social capital, in turn, reflects the ability of a local community to sustain a strong cohesion based on explicit norms, high levels of trust, and dynamic networks with a wide set of stakeholders (Gutiérrez et al. 2011) and can strengthen the

Fig. 7.2 Schematic showing (**a**) important social and institutional principles and (**b**) intrinsic values from biodiversity assets to ensure the expected outcomes (**c**) in community-based management (CBM) of biodiversity assets

management model as a whole (Pretty 2003). Therefore, identifying communities with strong local leaders and social capital should be a starting point in the process of CBM establishment.

The use of culturally noteworthy species as a flagship species is another important strategy, which can improve the engagement of local people in conservation and management initiatives (Garibaldi and Turner 2004; Freitas et al. 2020). Culturally important species play a central role in community subsistence, material acquisition, medicine, cultural identity, and/or spiritual values (Cristancho and Vining 2004; Garibaldi and Turner 2004). In Amazonia, there are a large number of species closely associated with human culture. For example, both arapaima and freshwater turtles have had a strong subsistence and cultural importance since pre-Columbian times (Prestes-Carneiro et al. 2016; Freitas et al. 2020). Besides this form of cultural importance, the economic value of the target species may also be important, in order to sustain a value chain and generate income for rural communities. Reliable economic returns can increase the likelihood of creating a cohesive management system, with high levels of engagement and compliance among users. However, to ensure sustainable harvests of a high-value species, the population size of the target species must be large enough to tolerate the harvesting dynamic (Ostrom 2009). For example, the management of arapaima, a culturally important species with high economic value and large population sizes within protected lakes, has generated substantial income at many sites across Amazonia (Campos-Silva et al. 2017). In contrast, freshwater turtles also have a high subsistence and market value, but do not currently generate a financial return, because in many localities the population size is not large enough to support a sustainable harvest and the commercial exploitation of these species is still illegal (Campos-Silva et al. 2018).

To ensure the sustainable harvesting of biodiversity assets, one of the most prominent characteristics of these Amazonian CBM models is the establishment of well-defined no-take zones (Campos-Silva et al. 2017), which arguably represents a cornerstone principle in common theory (Ostrom 2009). Explicit zoning of harvests, including "no-take" areas between human settlements, may ensure the successful reproduction of target species and the replenishment of wild populations through source-sink dynamics (Novaro et al. 2000; Levi et al. 2009; Antunes et al. 2016). This is highly relevant in Amazonian floodplains, where many fish species can move between aquatic environments during the flood pulse (Junk et al. 1989). In this context, it is also important to take into account the ecological requirements of the target species, such as their life cycle, habitat preferences, and migration behavior, to ensure that suitable habitats for foraging and reproduction are included in the spatial arrangement of the management zones (Campos-Silva et al. 2019). Spatial zoning appears to be best enforced, at least in the case of arapaima and freshwater turtles, through a strong surveillance system conducted by local residents, which ensures compliance and precludes poaching.

A clearly defined harvest quota is another essential component of sustainable harvest program (Costello et al. 2008). In the case of arapaima, the Brazilian government allows up to 30% of the adult population to be harvested, conditioned to some organizational requirements, including suitable infrastructure, a well-established value chain, and appropriate community organization (Campos-Silva and Peres 2016). While freshwater turtles are currently not legally harvested to generate income, there is already a trial program to work toward a sustainable quota for harvesting hatchlings, which can then be reared in captivity and later sold in local markets (Alho 1985; Andrade 2007).

Institutional principles may also be strong predictors of effective co-management initiatives (Berkes 2007; Ostrom 2009), ensuring a high level of autonomy, decentralizing the decision-making process, and reducing the overall costs of conservation (Somanathan et al. 2009). Respecting sociocultural contexts and Local Ecological Knowledge (LEK), for example, is an important component of successfully common pool resources management (Baggio et al. 2016). In the arapaima CBM, fishers' knowledge is one of the most important attributes sustaining the entire management system (Castello et al. 2011). In particular, participatory monitoring provides an opportunity to include LEK in management arrangements, and strongly contributes to the process of empowering local communities (Constantino et al. 2012). Both case studies we presented here exemplify how important LEK is for generating relevant information, such as comprehensive population assessments (Castello et al. 2009; Campos-Silva and Peres 2016; Campos-Silva et al. 2018).

Explicit regulations are important to ensure governance transparency, which is also a very important component of an effective natural resource management (Lockwood et al. 2010). Governments can play an important role regulating the management activity, such as in the case of arapaima, where the quota and permits are authorized by the Brazilian federal government. Adaptive management, which covers many others principles, including the use of LEK to evaluate and respond to internal forcing from the environment (Berkes et al. 2000), increases the speed of local responses to unforeseen circumstances, improving the resilience of the activity (Olsson et al. 2004).

Social rights and local aspirations of indigenous and nonindigenous dwellers in rural Amazonia have been neglected for a long time by the Brazilian government, compromising local autonomy and capacity building (Schwartzman et al. 2010; Vadjunec et al. 2011). To redress this balance, multiscale partnerships and subsidies are fundamental principles for the management of Amazonian common pool resources (Berkes 2007; Ostrom 2009), and are strongly desired in the initial stage of CBM establishment. The combination of different skills from multiple institutions, aligned with funding to boost management practices, can help to break the inertia built up over the past centuries in terms of local participation in decision making and the lack of technical expertise often observed in rural Amazonia.

The Amazon has been a scenario of profound interactions between humans and wildlife for millennia (see Clement et al. Chap. 3). This strong human–nature relationship has created a rich knowledge bank, used by human civilizations to build their social–ecological systems (Odling-Smee et al. 2003; Albuquerque et al. 2019). However, over the last 30 years, industrialized Brazilian society has destroyed more than 436,000 km^2 of Amazonian forest (INPE 2019), dramatically impacting local livelihoods and biodiversity. The alignment of biodiversity conservation and local welfare is one of the most imperative needs in Brazil today, both for biodiversity conservation and for the social justice of traditional tribal and nontribal populations still inhabiting the Amazon. Focusing on Amazonian bright spots can help us send a clear message of hope and action, which is critical to awaken the attention of local to international policymakers, and encourage both managers and stakeholders to increase their efforts to implement these successful management examples elsewhere.

References

Albuquerque UP, de Medeiros PM, Ferreira Júnior WS et al (2019) Social-ecological theory of maximization: basic concepts and two initial models. Biol Theory. https://doi.org/10.1007/s13752-019-00316-8

Alho CJR (1985) Conservation and management strategies for commonly exploited amazonian turtles. Biol Conserv 32(4):291–298. https://doi.org/10.1016/0006-3207(85)90019-9

Alves-Pinto HN, Hawes JE, Newton P et al. (2018) Economic impacts of payments for environmental services on livelihoods of agroextractivist communities in the Brazilian Amazon. Ecol Econ. 152:378–88. https://doi.org/10.1016/j.ecolecon.2018.05.016

Andrade PCM (2007) Criação e Manejo de Quelônios no Amazonas – Projeto Diagnóstico da Criação Animais Silvestres no Estado do Amazonas, 2nd edn. ProVárzea/FAPEAM/SDS, Manaus

Antunes AP, Fewster R, Venticinque EM et al (2016) Empty forest or empty rivers? A century of commercial hunting in Amazonia. Sci Adv 2(10):e1600936. https://doi.org/10.1126/sciadv.1600936

Arantes ML, Freitas CEC (2016) Effects of fisheries zoning and environmental characteristics on population parameters of the tambaqui (Colossoma macropomum) in managed floodplain lakes in the Central Amazon. Fish Manag Ecol 23(2):133–143. https://doi.org/10.1111/fme.12164

Arias L, Barbieri C, Barreto G et al (2018) High-resolution mitochondrial DNA analysis sheds light on human diversity, cultural interactions, and population mobility in Northwestern Amazonia. Am J Phys Anthropol 165(2): 238–255. https://doi.org/10.1002/ajpa.23345

Baggio JA, Barnett AJ, Perez-Ibara I et al (2016) Explaining success and failure in the commons: the configural nature of Ostrom's institutional design principles. Int J Commons 10:417–439. https://doi.org/10.18352/ijc.634

Berkes F (2007) Community-based conservation in a globalized world. Proc Natl Acad Sci 104(39):15188–15193. https://doi.org/10.1073/pnas.0702098104

Berkes F, Colding J, Folke C (2000) Rediscovery of traditional ecological knowledge as adaptive management. Ecol Appl 10:1251–1262. https://doi.org/10.2307/2641280

Brauner CJ (2004) Transition in organ function during the evolution of air-breathing; insights from Arapaima gigas, an obligate air-breathing teleost from the Amazon. J Exp Biol 207:1433–1438. https://doi.org/10.1242/jeb.00887

Bruner AG, Gullison RE, Rice RE, Da Fonseca GAB (2001) Effectiveness of parks in protecting tropical biodiversity. Science 291(5591):125–128. https://doi.org/10.1126/science.291.5501.125

Bruner AG, Gullison RE, Balmford A (2004) Financial costs and shortfalls of managing and expanding protected-area systems in developing countries. Bioscience 54(12):1119–1126. https://doi.org/10.1641/0006-3568(2004)054[1119:FCASOM]2.0.CO;2

Campos-Silva JV, Peres CA (2016) Community-based management induces rapid recovery of a high-value tropical freshwater fishery. Sci Rep 6:34745. https://doi.org/10.1038/srep34745

Campos-Silva JV, da Fonseca Junior SF, da Silva Peres CA (2015) Policy reversals do not bode well for conservation in Brazilian Amazonia. Nat e Conserv 13:193–195. https://doi.org/10.1016/j.ncon.2015.11.006

Campos-Silva JV, Peres CA, Antunes AP et al (2017) Community-based population recovery of overexploited Amazonian wildlife. Perspect Ecol Conserv 15:266–270. https://doi.org/10.1016/j.pecon.2017.08.004

Campos-Silva JV, Hawes JE, Andrade PCM, Peres CA (2018) Unintended multispecies co-benefits of an Amazonian community-based conservation programme. Nat Sustain 1:650–656. https://doi.org/10.1038/s41893-018-0170-5

Campos-Silva JV, Hawes JE, Peres CA (2019) Population recovery, seasonal site fidelity and daily activity of pirarucu (Arapaima spp.) in an Amazonian floodplain mosaic. Freshw Biol 64:1255–1264. https://doi.org/10.1111/fwb.13301

Cantarelli VH, Malvasio A, Verdade LM (2014) Brazil's Podocnemis expansa conservation program: retrospective and future directions. Chelonian Conserv Biol 13:124–128. https://doi.org/10.2744/ccb-0926.1

Castello L (2004) A method to count pirarucu Arapaima gigas: fishers, assessment, and management. N Am J Fish Manag *24*(2):379–389. https://doi.org/10.1577/M02-024.1

Castello L, Stewart DJ (2010) Assessing CITES non-detriment findings procedures for Arapaima in Brazil. J Appl Ichthyol 26:49–56. https://doi.org/10.1111/j.1439-0426.2009.01355.x

Castello L, Viana JP, Watkins G et al (2009) Lessons from integrating fishers of Arapaima in small-scale fisheries Management at the Mamirauá Reserve, Amazon. Environ Manag 43:197–209. https://doi.org/10.1007/s00267-008-9220-5

Castello L, Viana JP, Pinedo-Vasquez M (2011) Participatory conservation and local knowledge in the Amazon várzea: the pirarucu management scheme in Mamirauá. In: Pinedo-Vasquez M, Ruffino ML, Padoch C, Brondízio ES (eds) The Amazon Várzea. Springer, Dordrecht, pp 259–273

Castello L, Arantes CC, Mcgrath DG et al (2015) Understanding fishing-induced extinctions in the Amazon. Aquat Conserv Mar Freshw Ecosyst 25(5):587–598. https://doi.org/10.1002/aqc.2491

Cavole LM, Arantes CC, Castello L (2015) How illegal are tropical small-scale fisheries? An estimate for arapaima in the Amazon. Fish Res 168:1–5. https://doi.org/10.1016/j.fishres.2015.03.012

Cinner JE, Daw TM, McClanahan TR et al (2012a) Transitions toward co-management: the process of marine resource management devolution in three east African countries. Glob Environ Chang 22:651–658. https://doi.org/10.1016/j.gloenvcha.2012.03.002

Cinner JE, McClanahan TR, MacNeil MA et al (2012b) Comanagement of coral reef social-ecological systems. Proc Natl Acad Sci 109:5219–5222. https://doi.org/10.1073/pnas.1121215109

Coetzee BWT, Gaston KJ, Chown SL (2014) Local scale comparisons of biodiversity as a test for global protected area ecological performance: a meta-analysis. PLoS One 9:e105824. https://doi.org/10.1371/journal.pone.0105824

Costello C, Gaines SD, Lynham J (2008) Can catch shares prevent fisheries collapse? Science 321(5896):1678–1681. https://doi.org/10.1126/science.1159478

Cox M, Arnold G, Tomás SV (2010) A review of design principles for community-based natural resource management. Ecol Soc 15:38. https://doi.org/10.5751/ES-03704-150438

Cox M, Villamayor-Tomas S, Arnold G (2016) Design principles in commons science: a response to "Ostrom, Hardin and the commons" (Araral). Environ Sci Pol 61:238–242. https://doi.org/10.1016/j.envsci.2016.03.020

Cristancho S, Vining J (2004) Culturally defined keystone species. Res Hum Ecol 11(2):153-164

Cvitanovic C, Hobday AJ (2018) Building optimism at the environmental science-policy-practice interface through the study of bright spots. Nat Commun 9:1–5. https://doi.org/10.1038/s41467-018-05977-w

de Almeida MB (2002) The politics of Amazonian conservation: the struggles of rubber tappers. J Lat Am Anthropol 7:170–219. https://doi.org/10.1525/jlca.2002.7.1.170
Constantino PAL, Carlos HSA, Ramalho EE et al (2012) Empowering local people through community-based resource monitoring: a comparison of Brazil and Namibia. Ecol Soc 17:22. https://doi.org/10.5751/ES-05164-170422
Fearnside PM (1989) Extractive reserves in Amazonia Brazilian: an opportunity to maintain tropical rain forest under sustainable use. Bioscience 39(6):387–393. https://doi.org/10.2307/1311068
Freitas CT, Lopes PFM, Campos-Silva JV, Noble MM, Dyball R, Peres CA (2020) Co-management of culturally important species: A tool to promote biodiversity conservation and human well-being. People Nat 2:61–81. https://doi.org/10.1002/pan3.10064
Garibaldi A, Turner N (2004) Cultural keystone species: implications for ecological conservation and restoration. Ecol Soc 9:1. https://doi.org/10.5751/ES-00669-090301
Gibson CC, Marks SA (1995) Transforming rural hunters into conservationists: an assessment of community-based wildlife management programs in Africa. World Dev 23:941–957. https://doi.org/10.1016/0305-750X(95)00025-8
Gutiérrez NL, Hilborn R, Defeo O (2011) Leadership, social capital and incentives promote successful fisheries. Nature 470:386–389. https://doi.org/10.1038/nature09689
Hansen MC, Potapov PV, Moore R et al (2013) High-resolution global maps of 21st-century forest cover change. Science 342(6160):850–853. https://doi.org/10.1126/science.1244693
Hawes JE, Peres CA, Riley LB, Hess LL (2012) Landscape-scale variation in structure and biomass of Amazonian seasonally flooded and unflooded forests. For Ecol Manag 281:163–176. https://doi.org/10.1016/j.foreco.2012.06.023
INPE (2019) Monitoramento da Floresta Amazônica Brasileira por satélite – Projeto Prodes. http://terrabrasilis.dpi.inpe.br/app/dashboard/deforestation/biomes/legal_amazon/rates. Accessed 16 Apr 2019
Jepson PR, Caldecott B, Schmitt SF et al (2017) Protected area asset stewardship. Biol Conserv 212:183–190. https://doi.org/10.1016/j.biocon.2017.03.032
Junk WJ, Bayley PB, Sparks RE (1989) The flood pulse concept in river-floodplain-systems. Can J Fish Aquat Sci 106:110–127
Levi T, Shepard GH, Ohl-Schacherer J et al (2009) Modelling the long-term sustainability of indigenous hunting in Manu National Park, Peru: landscape-scale management implications for Amazonia. J Appl Ecol 46:804–814. https://doi.org/10.1111/j.1365-2664.2009.01661.x
Lima D, Pozzobon J (2005) Amazônia socioambiental: sustentabilidade ecológica e diversidade social. Estud Avançados 19:45–76. https://doi.org/10.1590/s0103-40142005000200004
Lockwood M, Davidson J, Stratford E, Griffith R (2010) Governance principles for natural resource management. Soc Nat Resour An Int J 23:986–1001. https://doi.org/10.1080/08941920802178214
Magnusson WE, Grelle CEV, Marques MCM et al (2018) Effects of Brazil's political crisis on the science needed for biodiversity conservation. Front Ecol Evol 6. https://doi.org/10.3389/fevo.2018.00163
Miorando PS, Rebêlo GH, Pignati MT, Brito Pezzuti JC (2013) Effects of community-based management on Amazon River turtles: a case study of Podocnemis sextuberculata in the lower Amazon floodplain, Pará, Brazil. Chelonian Conserv Biol 12:143–150. https://doi.org/10.2744/ccb-1011.1
Muehlig-hofmann A (2007) Traditional authority and community leadership: key factors in community-based marine resource management and conservation. SPC Tradit Mar Resour Manag Knowl Inf Bull 21:31–44
Nelson JS (1994) Fishes of the world. Wiley, New York. https://doi.org/10.1002/9781119174844
Newton P, Watkinson AR, Peres CA (2011) Determinants of yield in a non-timber forest product: Copaifera oleoresin in Amazonian extractive reserves. For Ecol Manag 261:255–264. https://doi.org/10.1016/j.foreco.2010.10.014
Novaro AJ, Redford KH, Bodmer RE (2000) Effect of hunting in source-sink systems in the neotropics. Conserv Biol 14:713–721. https://doi.org/10.1046/j.1523-1739.2000.98452.x

Odling-Smee FJ, Laland KN, Feldman MW (2003) Niche construction: the neglected process in evolution. Princeton University Press, Princeton
Olsson P, Folke C, Hahn T (2004) Social-ecological transformation for ecosystem management: the development of adaptive co-management of a wetland landscape in southern Sweden. Ecol Soc 9. https://doi.org/10.5751/ES-00683-090402
Ostrom E (2009) A general framework for analyzing sustainability of social-ecological systems. Science 325(5939):419–422. https://doi.org/10.1126/science.1172133
Peres CA (2011) Conservation in sustainable-use tropical forest reserves. Conserv Biol 25:1124–1129. https://doi.org/10.1111/j.1523-1739.2011.01770.x
Petersen TA, Brum SM, Rossoni F et al (2016) Recovery of Arapaima sp. populations by community-based management in floodplains of the Purus River, Amazon. J Fish Biol 89:241–248. https://doi.org/10.1111/jfb.12968
Pezzuti JCB, Lima JP, da Silva DF, Begossi A (2010) Uses and taboos of turtles and tortoises along Rio Negro, Amazon Basin. J Ethnobiol 30:153–168. https://doi.org/10.2993/0278-0771-30.1.153
Pinkerton E, Weinstein M (1995) Fisheries that work: sustainability through community-based management. The David Suzuki Foundation, Vancouver
Prestes-Carneiro G, Béarez P, Bailon S et al (2016) Subsistence fishery at Hatahara (750–1230 CE), a pre-Columbian central Amazonian village. J Archaeol Sci Reports 8:454–462. https://doi.org/10.1016/j.jasrep.2015.10.033
Pretty J (2003) Social capital and the collective management of resources. Science 302(5652): 1912–1914. https://doi.org/10.1126/science.1090847
Rebêlo G, Pezzuti J (2000) Percepções sobre o consumo de quelônios na Amazônia: sustentabilidade e alternativas ao manejo atual. Ambient Soc 6:85–104. https://doi.org/10.1590/S1414-753X2000000100005
Schwartzman S, Alencar A, Zarin H, Souza APS (2010) Social movements and large-scale tropical forest protection on the Amazon frontier: Conservation from chaos. J Environ Dev 19:274–299. https://doi.org/10.1177/1070496510367627
Schneider L, Ferrara CR, Vogt RC, Burger J (2011) History of turtle exploitation and management techniques to conserve turtles in the Rio Negro Basin of the Brazilian Amazon. Chelonian Conserv Biol 10:149–157. https://doi.org/10.2744/ccb-0848.1
Smith NJH (1979) Aquatic turtles of Amazonia: an endangered resource. Biol Conserv 16:165–176. https://doi.org/10.1016/0006-3207(79)90019-3
Somanathan E, Prabhakar R, Singh B (2009) Decentralization for cost-effective conservation. Proc Natl Acad Sci 106:4143–4147. https://doi.org/10.1007/3-540-45701-1
Sunderlin WD, Angelsen A, Belcher B et al (2005) Livelihoods, forests, and conservation in developing countries: an overview. World Dev 33:1383–1402. https://doi.org/10.1016/j.worlddev.2004.10.004
Terborgh J, Peres CA (2017) Do community-managed forests work? A biodiversity perspective. Land 6:1–7. https://doi.org/10.3390/land6020022
Vadjunec JM, Schmink M, Greiner AL (2011) New Amazonian geographies: emerging identities and landscapes. J Cult Geogr 28:1–20. https://doi.org/10.1080/08873631.2011.548477
Watson JEM, Dudley N, Segan DB, Hockings M (2014) The performance and potential of protected areas. Nature 515:67–73. https://doi.org/10.1038/nature13947

Chapter 8
Productive Restoration as a Tool for Socioecological Landscape Conservation: The Case of "La Montaña" in Guerrero, Mexico

Eliane Ceccon

8.1 Introduction

Mexico is home to 12% of the world's biodiversity, with 200,000 different species (CONABIO 2018). However, approximately 50% of Mexican territory has some degree of degradation (Bollo-Manent et al. 2014), associated with difficult social issues. Around 56.3% of the population lives in poverty, 20.1% lack access to food, and 57.2% of jobs are nonformal (SEDESOL 2018). A large percentage of the population is indigenous, though this varies among states, from 0.22% in Zacatecas to 58.96% in Yucatán (INEE 2005). In 2007, collectively owned lands (*ejidos* or communities) accounted for 53.4% of Mexican territory (INEGI 2007), home to 3,448,470 people (50% of whom presented high marginalization indices; Bray et al. 2007). These common-pool systems also prevail in 60% of Natural Protected Areas (Bezaury-Creel and Gutiérrez-Carbonell 2009).

The serious deterioration of Mexico's ecosystems and its effects on many different vulnerable social groups highlights the need to develop strategies for conservation and restoration that can prevent the reduction of biological diversity and promote the restoration of ecosystem services as well as quality of life and well-being of local people (Ramos 2005; Ceccon 2013; Ceccon et al. 2015). Thus, in Mexico, restoration activities depend on strategies that go beyond those normally used in the principles of the International Society of Ecological Restoration (SER 2004; Ceccon and Perez 2017); they must follow an interdisciplinary approach, with close coordination and cooperation with the affected social group. Forest restoration has the additional objective of guaranteeing the health of the landscape to provide economic benefits in the medium- and long term (Ceccon 2013). Another

E. Ceccon (✉)
Centro Regional de Investigaciones Multidisciplinarias, Universidad Nacional Autónoma de México, Cuernavaca, Mexico

C. Baldauf (ed.), *Participatory Biodiversity Conservation*,
https://doi.org/10.1007/978-3-030-41686-7_8

important consideration is that in Mexico, the vast majority of ecosystem restoration projects are publicly financed, making it particularly important to uphold the public trust by using those funds responsibly and effectively (Méndez-Toribio et al. 2018).

The "La Montaña" region of Guerrero State is considered one of the poorest and least developed areas in Mexico (CONEVAL 2012), with a Human Development Index (HDI) of 0.515 (CDI 2000; Sierra 2007). The region does not have access to health services, schools, paved roads, telecommunications, or electricity, and its situation has been further exacerbated by the recent increase in militarization, social conflict, and violence (Camacho 2007; Gaussens 2018). Around 72% of the terrain in this region is steeply sloped (INEGI 2010), highly fragmented, and visibly degraded (Miramontes et al. 2012; Borda-Niño et al. 2017a).

At same time, "La Montaña" is also highly culturally diverse, including four indigenous groups (Mixtecos, Nahuas, Amuzgos, and Me'Phaas), and a long-standing tradition of collective actions, which has generated a large number of civil society organizations (CDI 2005).

The nongovernmental organization (NGO) Xuajin Me'Phaa was created in 2006 to integrate families dedicated to the organic production of honey, beans, bananas, pineapple, coffee, and hibiscus (*Hibiscus sabdariffa*). Its board of directors is elected by the 300 producers. Of these, 124 are active producers of organic hibiscus, which is the most economically important among the products, because it is sold to a national supermarket chain. Originally, the hibiscus was sold to the supermarket as part of a charitable program in which all revenues went to the NGO (Galicia-Gallardo et al. 2018), but their participation in that program later ended, and they are currently regular suppliers of organic hibiscus for the supermarket chain's 126 stores. Xuajin Me'Phaa's organic products are also sold in small organic street markets (Hernández-Muciño et al. 2018).

At the same time, the Regional Center for Multidisciplinary Research (CRIM, in Spanish) of the National Autonomous University of Mexico (UNAM, in Spanish), began working with Xuajin Me'Phaa in 2008, and in 2013 signed an agreement to establish several restoration projects, with active participation of the members of the organization. The work of this research group employs the participatory action research (PAR) methodology. In this case, the members of the organization participate as coresearchers and learning partners (McNiff 2013). The NGO members are the protagonists of the research, interacting throughout the research process through design, actions, and proposals, among other aspects (Alberich-Nistal 2008). This process leads to "social learning," a term that has been used to refer to the processes of learning and changes in the relationship between individuals and ecological systems (Muro and Jeffrey 2008). The ultimate objective is for the organization to become more empowered in their use of natural resources, take ownership of the tools generated by the research in which they have participated, and develop their own independent restoration projects.

In the socioecological reality of Xuajin Me'phaa, the projects suggested by the research group must improve not only ecosystem services, but also the supply of goods such as wood and food. In this sense, "productive restoration" was the main restoration concept behind the projects. This concept refers to the restoration of some elements of the structure and function of the original ecosystem, together with the productivity of the land, in a sustainable way, using agroforestry and agroecological tools, with the aim of offering products that generate economic benefits to the local population (Ceccon 2013).

The first step in this process was an attempt to diagnose the state of degradation and fragmentation of the three micro-basins of the Acatepec municipality, where the majority of the lands of the members of the NGO are located. This analysis used tools from Geographic Information Systems. We also did a phytosociological analysis of the reference ecosystems (SER 2004). The next step was to characterize the consumption of natural resources (mainly fuelwood) and identify the species most frequently used for this purpose. Another important aim was to evaluate the sustainability of organic hibiscus production in order to identify the critical points of this system. Finally, productive restoration experiments were also established to increase crop yields and sustainability.

In this chapter we also describe the most important results of the participatory action research carried out in "La Montaña"—a restoration project established 5 years ago, whose financing and management was carried out by the NGO itself to restore 200 traditional homegardens (THs). The motivations behind the projects are presented, as well as their potential role in landscape connectivity and biodiversity conservation.

8.2 Description of the Study Site

Most of the members of the NGO live in the Acatepec municipality of Guerrero State, between 17°00′ and 17°22′ north latitude and 98°49′ and 99°11′ west longitude, at an altitude between 300 and 2600 m.a.s.l. (INEGI 2010). The climate of the region is warm and subhumid, with a total annual rainfall of ~1800 mm. The distribution of the precipitation is seasonal; the rainy season lasts from April to November, with highest rainfall in September (434 mm). The average annual temperature is 25.7 °C; May is the warmest month and January is the coldest (SMN 2013). This territory is predominantly agricultural with a rough topography; 70% of the total surface is steeply sloped, 20% is semi-flat, and only 10% is flat (INEGI 2010). The predominant soil type is Umbric Stagnic Fluvisol (Episkeletic, clay), which is characterized by being shallow and not very developed, formed by alluvial materials deposited in terraces (WRB 2007). The vegetation is tropical deciduous forest in the low-lying areas and coniferous-oak forest at higher altitudes (Miranda and Hernández 1963; Borda-Niño et al. 2017a).

8.3 Diagnostic Phase

8.3.1 Level of Fragmentation and Reference Ecosystems

In this phase, we sought to describe the spatial distribution of remnants of native vegetation in the landscape of three micro-basins of the Acatepec municipality (total area 135 km^2), and to characterize the vegetation at three altitudes between 520 and 2600 m.a.s.l. to establish reference ecosystems for this region.

We found that the landscape of the three micro-basins was strongly fragmented. A significant proportion of the fragments above 1000 m.a.s.l. altitude exhibited edge effect due to their irregular shapes and small sizes (<21 ha). The situation of the tropical deciduous forest vegetation below 1000 m.a.s.l. was even more critical, since most of the fragments were very small (<10 ha) and irregularly shaped. Moreover, the majority of the fragments were separated by more than 1 km, leading to particularly severe effects on species with low dispersal capacity (Figs. 8.1 and 8.2; Borda-Niño et al. 2017a). The phytosociological analysis of the reference ecosystems showed that the native vegetation of the highest-altitude zones (2606–1072 m) was composed of mixed forests dominated by species of the *Quercus* and *Pinus* genera of boreal affinity. The most important species (according to the Relative Importance Index) were *Quercus scytophylla*, *Quercus obtusata*, *Quercus elliptica*, *Pinus maximinoi*, *Pinus oocarpa*, and *Clethra lanata*. On the other hand, the native vegetation of the lowest-altitude zones (520–1071 m.a.s.l.) was composed of tropical deciduous forest species. The most important species were *Pseudobombax ellipticum*, *Cochlospermum vitifolium*, *Spondias purpurea*, *Comocladia macrophylla*, and *Quercus glaucescens* (Fig. 8.2; Borda-Niño et al. 2017a).

8.3.2 Fuelwood Consumption Diagnosis

Through 60 surveys at the same three altitude ranges, in the three micro-basins of the Municipality of Acatepec, we found that all of the households surveyed used fuelwood daily, with an average consumption of between 1.70 kg and 2.06 kg per person per day. Nearly half (45%) of the users collected fuelwood from standing trees, mainly from *Quercus* species (details in Salgado et al. 2017). Considering an average consumption per person of 1.59 kg (±0.42) of biomass, a population of around 1837 people, and the fact that 45% of fuelwood is extracted from standing trees, around 492,655 kg of fuelwood biomass is extracted per year.

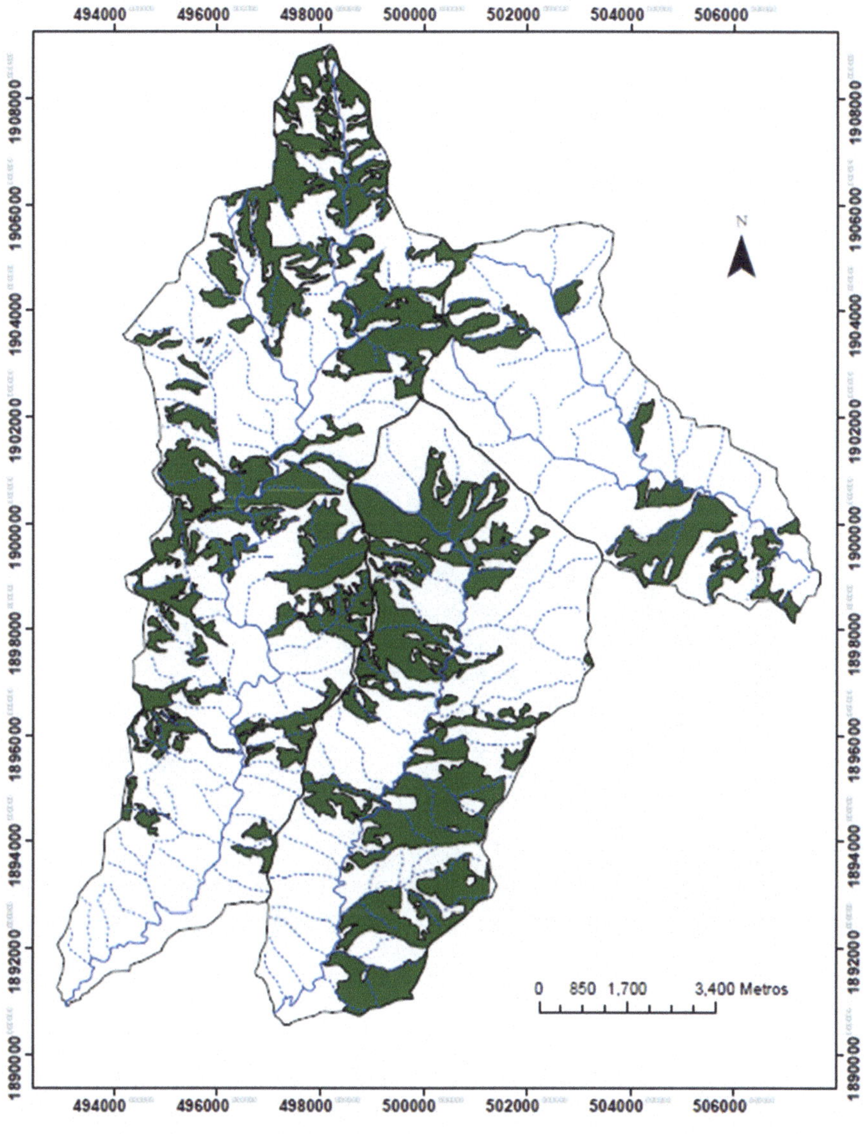

Fig. 8.1 Distribution of the fragments in the analysis of functional connectivity of the landscape in three micro-basins of the Municipality of Acatepec, Guerrero. (*Source*: Borda-Niño 2014)

8.3.3 Sustainability of Organic Hibiscus Production

To evaluate the sustainability of hibiscus productions by Xuajin Me'Phaa members, we used the "Framework for the Evaluation of Natural Resource Management Systems Incorporating Sustainability Indicators" (MESMIS, in Spanish), which

Fig. 8.2 Remnants of tropical deciduous forest in the lowest altitude of three micro-basins of the Municipality of Acatepec, Guerrero. (*Photo*: Monica Borda Niño)

allows the comparison between reference (conventional) and alternative (organic) agroecosystems (López-Ridaura et al. 2002). This process is cyclical, and certain steps are necessary to define the dimensions, attributes, and criteria to measure the indicators of their evaluation units and reference or ideal values. Finally, the communication of results is fundamental to provide a feedback to management decisions (López-Ridaura et al. 2005).

Organic production of hibiscus showed high levels of sustainability. The main critical point was hibiscus yield. On the other hand, the organic production had high levels of soil organic matter and erosion control, as well as a high benefit/cost ratio due to the premium prices offered in the organic product market. The social capital index in the NGO also scored highly, considering aspects such as training, trust, organization, participation, and the capacity to change and to resolve conflicts (Galicia-Gallardo et al. 2019). This was a fundamental result, since social capital is recognized as one of the aspects of a community that is most likely to promote local development (Castelletti and Canzanelli 2005).

8.4 Experimental Phase

8.4.1 Amendments Used to Increase Organic Hibiscus Yield

Because the sustainability analysis of organic hibiscus cultivation mentioned above, revealed a lower-than-optimal hibiscus yield, we tested five amendments (and some of their combinations) to increase this yield. The amendments used in the hibiscus plantation were:

- Detritus remaining after the hibiscus harvest (H)—normally used by the organic hibiscus farmers.
- Leaves of *Calliandra houstoniana* (C)—native leguminous species, never used by organic hibiscus farmers, but recognized by them as soil improver.
- *Mucuna pruriens* (M)—green manure widely used by the organic hibiscus farmers.
- Biofertilizer (B)—consisting of a mixture of *Azospirillum brasilense* (AzoFer®, BioFabrica Siglo XXI, 500 million CFU/gram of fertilizer) and *Rhizophagus irregularis* (MicorrizaFer®, BioFabrica Siglo XXI, 100,000 propagules/gram of fertilizer) in sterile medium (soil-peat), and never used by organic hibiscus farmers.

The amendments were combined in the following treatments: H + CH, H + CH + B, H + CH + M, H, H + B, and H + M. These treatments were distributed in a randomized block design in the alleys between planted hibiscus and analyzed by a one-way analysis of variance (ANOVA) and Duncan *pos hoc* test.

After 1 year, there were significant ($p < 0.05$) differences in the hibiscus yield among amendment treatments, with significantly higher yields in all of the treatments that included *C. houstoniana* compared to all of the treatments that did not (Fig. 8.3). One explanation for this result is the contribution of nutrients released by this species. Of the three species used in the forest agroecological system, *C. houstoniana* had the highest percentage of nitrogen (2.03 ± 0.015) in its leaves. Since *M. pruriens* and *H. sabdariffa* are already used by the organic farmers, the addition of *C. houstoniana* trees in an alley cropping system could improve the yield of the hibiscus as well as improving landscape connectivity (Silva Galicia, PhD Dissertation in preparation).

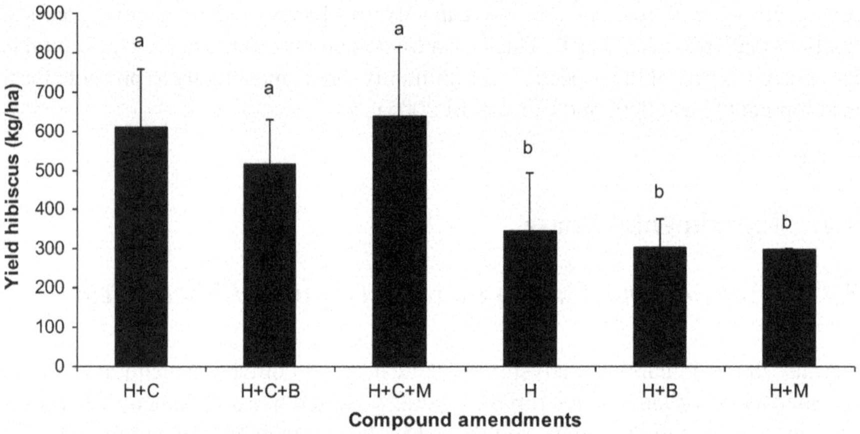

Fig. 8.3 Average yield of hibiscus (kg/ha) obtained in 2016 under different combined amendment treatments: C—Calliandra houstoniana; M—Mucuna pruriens; H—Hibiscus sabdariffa; B—biofertilizer. Different letters above the bars indicate significant differences ($p < 0.05$). Error bars show standard error. (*Source*: Silva-Galicia, PhD Dissertation in preparation)

8.4.2 The Role of Leucaena macrophylla *in Productive Restoration*

Leucaena macrophylla is a tree native to southern Mexico's tropical deciduous forest and belongs to a well-known tropical genus used in agroforestry systems. While this species is known and appreciated by people from "La Montaña," many of its benefits remain unexplored.

In this study, we found that *L. macrophylla* leaves contained a high percentage of nitrogen (3.52 ± 0.02) and its litter decomposed relatively rapidly (50% of mass lost over the first 6 months). This species also presented a high firewood value index (FVI = 2594.65), suggesting its high potential as a fuelwood. *L. macrophylla* also presented acceptable values as fodder; though a decomposition experiment showed higher-than-ideal lignin content, its in vitro digestibility remained high (57.76%). Thus, this species is also a promising source of livestock fodder (see details in Hernandez-Muciño et al. 2015). Thus, *L. macrophylla* is a promising species for use in productive restoration projects.

8.5 Results of Participatory Action Research (PAR): The Restoration of Traditional Homegardens

Participation is the main characteristic of action research, which involves a particular kind of interpersonal relationship that fades the boundaries between traditional roles of the researchers and the researched (Reason and Bradbury 2001). In the case of the work carried out between Xuajin Me'Phaa and the CRIM, participatory action research (PAR) refers to the relationship mentioned by Hall (2001) and Park (2001): the meeting of two groups—"researchers" on the one hand and "the community" on the other. The main objectives of our PAR project were to empower the community by helping them become aware of their own resources, increasing their problem-solving capacity, and helping them becoming more self-reliant and less dependent on external support (Fals-Borda 2001; Park 2001; Swantz et al. 2001). Our PAR approach also promoted "social learning," emphasizing social interactions among stakeholders, individual and group reflection on what is being learned, and iterative attempts to apply what is being learned to the issues under discussion (Pahl-Wostl and Hare 2004). The members of Xuajin Me'Phaa who participated in the management of these socioecological systems were also able to improve their adaptability by participating in decision-making processes (Folke et al. 2005; Fazey et al. 2017). In this case, learning was the result of active social participation in the practices of the community (Lave and Wenger 1991; Wenger 1998).

As a result of their active participation in the academic projects, in 2013, after 5 years of collaboration with CRIM in research projects of productive restoration, Xuajin Me'Phaa obtained its own financing to carry out productive restoration in practice. The project "Mbaá Yuskha: Me'Phaa Cultural Homegarden" obtained funding from the Ministry of Agriculture and Rural Development (SAGARPA, in Spanish) through a program linked with Civil Society Organizations. This project aimed to restore species among 200 traditional homegardens in the backyards of the community members. Mbaá Yuskha means "The Forest of the Grandparents" in Me'Phaa. This name refers to the fact that before the arrival of the Spaniards in Mexico, this ethnic group was nomadic, and the homegardens were planted to offer food to any group that passed through the place, for many years, even after the people who originally planted them had moved on.

In fact, in Mexico, homegardens are the product of a long history of management since the pre-Columbian period (Heindorf 2011) and are known as traditional (THs) or cultural homegardens. THs are spaces for production, selection, domestication, diversification, and conservation of flora and fauna that are closely related to the preservation and enrichment of cultural values and the generation and appropriation of technologies (Montañez-Escalante et al. 2012). They are also reservoirs of agrodiversity and biodiversity that protect plant species present in natural forest fragments and favor gene flow between the surrounding forests and the productive units. They also provide the population with multiple products to satisfy economic, social, and cultural needs (Montagnini 2006).

In the first phase of the project, community technicians were trained in soil and water conservation, livestock and agroforestry systems management, and the production of vegetables in micro-tunnels. In the second phase, THs were characterized in terms of plant species richness, number and type of domestic animals involved, source and availability of water for irrigation, and the community members' perception of soil fertility to design the intervention strategies. Subsequently, native and introduced plants that were chosen by the local inhabitants for inclusion in THs, agricultural tools, infrastructure material for irrigation, and materials for building livestock enclosures were distributed (Borda-Niño et al. 2017b).

In analyzing the motivations for TH restoration, the main priority was food security (47% of planted species), followed by cultural motivations (e.g., providing decoration for the house, church, religious festivities; 18% of species); however, 28% mentioned species had multiple purposes. Native (autochthonous) species were preferred (57%) over introduced (allochthonous) species (43%), and trees (52%) were preferred over herbaceous and shrub species (31% and 17%, respectively). Most of these species reported were pollinated (90%) and dispersed (64%) by animals, indicating a potential role of these agroforestry systems in landscape conservation (Aguirre-Salcedo 2018).

In terms of animal husbandry, 90% of THs housed chickens, 50% pigs, and 17% goats. Goats and pigs are consumed just once or twice a year (Aguirre-Salcedo 2018). In "La Montaña," natural resources for cattle ranching are scarce, since 72% of the terrain is steeply sloped and water is scarce (INEGI 2010), so the absence of cattle is not surprising.

The average size of the THs was 464.5 ± 59 m^2 (~0.05 ha; Aguirre-Salcedo 2018), among the smallest reported in Mexico (Alvarez-Buylla et al. 1989; Rico-Gray et al. 1990; Clerck and Negreros-Castillo 2000). A total of 3509 individuals and 141 species were found, with an average of 117 ± 16 individuals and 23 ± 1 species per TH. In terms of species richness, this is seventh among ten homegardens around the world evaluated by Abebe (2005). The low number of species may be due to the small size of the THs and water scarcity. The species with the highest relative importance percentage were banana (*Musa* ssp., 55%), mango (*Mangifera indica*, 16%), pineapple (*Ananas Comosus*, 10%) and papaya (*Carica papaya*). Three of these most important species appeared in the TH of Mr. José Nava, who is the representative of the THs established through the Xuajin Me'Phaa TH restoration project (Fig. 8.4). The diversity index (H´) in the THs varied between 2.02 (1430 m.a.s.l.) in Xochitepec and 2.48 in Alcamani (1269 m.a.s.l.) and Naranjo (781 m.a.s.l.).

As homegardens are recognized as a more productive, sustainable, and appropriate food production system to the reality of small farmers (Farrell and Altieri 1999), there is a recent interest in looking at this type of management as a biodiversity conservation alternative (Tscharntke et al. 2005; Melo et al. 2013).

At the same time, Mexico has the eleventh highest total bird richness and fourth highest proportion of endemic bird taxa worldwide (Navarro-Sigüenza et al. 2014). The level of fragmentation in the study area could threaten this diversity. We therefore sought to evaluate the ecological importance of THs for the conservation of the bird community in one locality of the region known as Plan de Gatica (538 m.a.s.l.). For this purpose, was evaluated the way in which birds used 15 THs, recording the activities they carried out within them.

In order to have a representative sample of the birds that visit the THs, sampling was carried out once during the rainy season and once during the dry season. In each TH, an observation point was established where the birds visited the TH. The activity they performed inside the TH and the stratum in which they spent most of their time during the visit were also recorded. Three vegetation strata were considered:

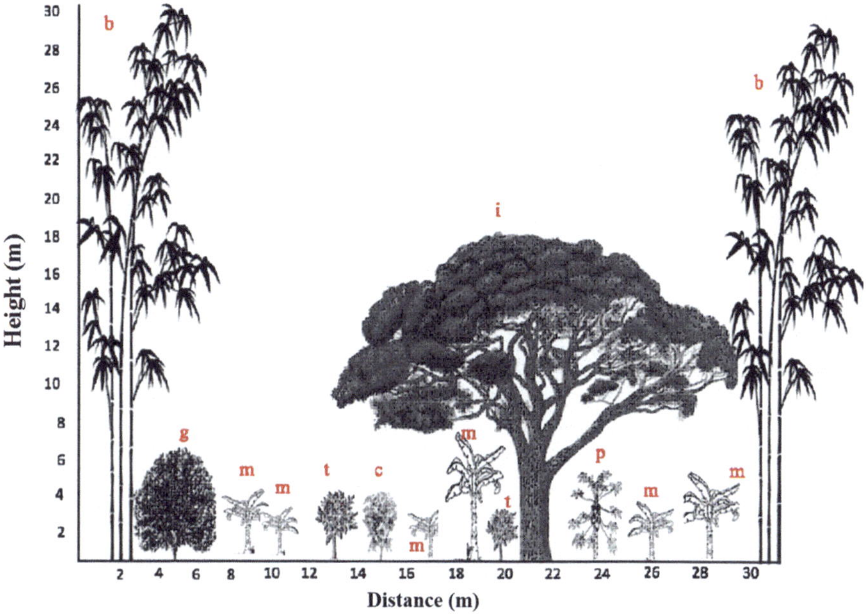

Fig. 8.4 Vertical profile of the traditional homegarden of Mr. José Nava in the town of Plan de Gatica: b—*Bambusa textiles* (bamboo); g—*Psidium guajava* (guava); m—*Musa* sp. (banana); t—*Theobroma cacao* (cocoa); c—*Coffea arabica* (coffee); p—*Carica papaya* (papaya); and i—*Mangifera indica* (mango). (*Source*: Aguirre-Salcedo 2018)

low (0–5 m high), medium (5–10 m high), and high (>10 m). Sampling was carried once from 7:00 to 10:00 in each TH.

There were 1027 visits from 36 species of birds, belonging to 13 trophic guilds according to the classification of González-Salazar et al. (2014). These species represent 13% of the resident birds reported for the state of Guerrero by Navarro-Siguenza (1998). Although apparently it is a low percentage, it is a considerable number of species, taking into account that the sampling was limited to a very small total area (4654 m^2), only within the THs. Of 13 trophic guilds, seven corresponded to the insectivore guild. Some presented more specialized behaviors such as bark excavation insectivores and bark pecking insectivores. The frugivorous species that were reported were representatives of two trophic guilds: frugivorous peckers of the soil in the lower part of the canopy and peeping frugivores of the upper part of the canopy. The frequency of frugivores was surprisingly low in the THs, given the high number/percentage of zoocoric plants (Aguirre-Salcedo 2018). It is possible that the high degree of anthropogenic degradation in this landscape (Borda-Niño et al. 2017a) is the main constraint on the presence of frugivores, rather than the characteristics of the THs. The omnivores also represented two trophic guilds: the ground-foraging omnivores (GFO) and the arboreal foraging omnivores (OFA). Granivores only belonged to one trophic guild, like the nectarivores (Vargas Cárdenas 2018). The presence of 13 trophic guilds is indicative of an important functional richness, since they represented 59% of the trophic guilds of the birds of Mexico (González-Salazar et al. 2014). This suggests that THs offer resources for a wide variety of birds.

Some bird species were present in all or most of the THs, while others occurred in only a few or even in a single TH. Only one of the species (*Passer domesticus*) was exotic and was recorded in only three THs. It is worth noting that in 14 of the 15 THs studied, active nests were found, belonging to 10 species of birds (Vargas Cárdenas 2018).

All strata of vegetation were widely visited by birds, suggesting that all available spaces of THs are used by different groups of birds. In addition, many birds only stopped at the THs, which may indicate that they use them to move among fragments or for refuge. Others were seen foraging, evidence that they are an important source of food, and the presence of nests in nearly all of the THs demonstrates that they provide nesting habitat for birds.

8.6 Conclusions

The diagnostic phase was very important to characterize the level of landscape degradation and the species in reference ecosystems, quantify the demand for fuelwood, and identify the critical points in the organic hibiscus production.

The experimental phase does not have an end point, since we are continually striving to solve the critical points in crop production. Currently, two native multipurpose species have been studied (*Leucaena leucocephala* and *Calliandra houstoniana*) and have shown promising results in productive restoration projects, mainly in alley cropping systems. We are also evaluating the effectiveness and quality of the amendments traditionally used by farmers in combination with *Calliandra houstoniana* and biofertilizer.

The use of participatory action research as a working strategy over the past 10 years has empowered Xuajin Me'Phaa members in terms of their resources and implemented social learning about the importance of productive restoration in the landscape, without neglecting their food security.

Acknowledgments This work was supported by the Consejo Nacional de Ciencia y Tecnología (CONACYT) (student fellowships) and Dirección General de Asuntos del Personal Académico (DGAPA), Universidad Nacional Autónoma de México (UNAM; PAPIIT IN300119). The author appreciates the invaluable support of NGO Xuajin Me'Phaa members (specially Margarita Muciño), and of her students in the realization of their graduate theses: Ana Silva-Galicia, Citlali Aguirre-Salcedo, Diego Hernandez-Muciño, Freddy Vargas-Cardenas, Monica Borda-Niño, and Paola Galicia-Gallardo.

References

Abebe T (2005) Diversity in homegarden agroforestry systems of southern Ethiopia. PhD Dissertation, Wageningen University, The Netherlands

Aguirre-Salcedo C (2018) Evaluación de las motivaciones culturales, económicas y ambientales en la restauración de los traspatios de la organización no gubernamental Xuajin Me 'Phaa en Guerrero y una propuesta de monitoreo. MSc Dissertation, Universidad Nacional Autónoma de Mexico

Alberich-Nistal T (2008) Lap, redes y mapas sociales: desde la investigación a la intervención social. Portularia 8(1):17

Alvarez-Buylla ME, Lazos-Chavero E, Garcia-Barrios JR (1989) Homegardens of humid tropical region in Southeast Mexico: an example of an agroforestry cropping system in a recently established community. Agrofor Syst 8:133–156

Bezaury-Creel J, Gutiérrez-Carbonell D (2009) Áreas naturales protegidas y desarrollo social en México. In: CONABIO (ed) Capital natural de México, vol II. Comisión Nacional para el Conocimiento y Uso de la Biodiversidad, México, D.F, pp 382–431

Bollo-Manent B, Santana JRH, Linares APM (2014) The state of the environment in Mexico. Cent Eur J Geosci 6:219–228. https://doi.org/10.2478/s13533-012-0172-1

Borda-Niño M (2014) Distribución espacial de los remanentes de vegetación nativa a nivel de Microcuenca en un sector del municipio de Acatepec (Estado de Guerrero): implicaciones en actividades de restauración a nivel de paisaje. MSc Dissertation, CIGA, UNAM

Borda-Niño M, Hernández-Muciño D, Ceccon E (2017a) Planning restoration in human-modified landscapes: new insights linking different scales. Appl Geography 83:118–129. https://doi.org/10.1016/j.bbr.2011.03.031

Borda-Niño M, Carranza Santiago M, Hernández-Muciño D, Muciño-Muciño M (2017b) Productive restoration in practice: the case of Me'Phaa indigenous communities in "La Montaña" of Guerrero, Mexico. In: Ceccon E, Pérez DR (eds) Beyond ecological restoration: social perspectives in Latin America and Caribean. Vazquez Mazzini Editores, Argentina, pp 247–255

Bray DB, Merino L, Barry D (eds.) (2007) Los bosques comunitarios de México. Manejo sustentable de paisajes forestales, INE–SEMARNAT, Mexico. http://www.ine.gob.mx/publicaciones/consultaPublicacion.html?id_pub=532

Camacho Z (2007) Montaña de Guerrero: pobreza y militarización. Contralínea 70(5):70. http://www.contralinea.com.mx/archivo/2007/enero/htm/montana_guerrero_militares.htm. Accessed 9 Nov 2018

Castelletti O, Canzanelli G (2005) Estrategias e instrumentos para el desarrollo local en la era de la globalización. Ópera 5:78–88

CDI (Comisión Nacional para el Desarrollo de los pueblos indígenas) (2000) Informe sobre el desarrollo humano de los pueblos indígenas de México 2006 online version, base 2000. http://www.cdi.gob.mx/idh/idhpi_sintesis_ejecutiva.pdf. Accessed 10 Nov 2018

CDI (Comisión Nacional para el Desarrollo de los pueblos indígenas) (2005) Desarrollo indígena en 50 municipios Comisión Nacional para el Desarrollo de los pueblos indígenas http://www.cdi.gob.mx/50municipios/50_municipios.pdf. Accessed 10 Nov 2018

Ceccon E (2013) Restauración en bosques tropicales: fundamentos ecológicos, prácticos y sociales. Ediciones Díaz de Santos/UNAM, Mexico

Ceccon E, Perez DR (coords) (2017) Beyond ecological restoration: social perspectives in Latin America and Caribean. Vazquez Mazzini Editores, Argentina

Ceccon E, Barrera-Cataño JI, Aronson J, Martínez-Garza C (2015) The socioecological complexity of ecological restoration in Mexico. Restor Ecol 23(4):331–336. https://doi.org/10.1111/rec.12228

Clerck FAJ, Negreros-Castillo P (2000) Plant species of traditional Mayan homegardens of Mexico as analogs for multistrata agroforests. Agrofor Syst 48:303–317

CONABIO (2018) Mexican biodiversity. http://www.biodiversidad.gob.mx/v_ingles/index.htm. Accessed 12 Dec 2018

CONEVAL (Consejo Nacional de Evaluación de la Política de Desarrollo Social) (2012) Informe de pobreza y evaluación en el estado de Guerrero. CONEVAL, México, D.F

Fals-Borda O (2001) Participatory action research in social theory. In: Reason P, Bradbury H (eds) Handbook of action research. SAGE, London, pp 27–37

Farrell JG, Altieri MA (1999) Sistemas agroforestales. In: Agroecología: Bases científicas para una agricultura sustentable. Editorial Nordan-Comunidad, Montevideo, p 338

Fazey I, Carmen E, Rao-Williams J, Hodgson A, Fraser J, Cox L, Scott D, Tabor P, Robeson D, Searle BA, Lyon C, Kenter J, Murray B (2017) Community resilience to climate change: outcomes of the scottish borders climate resilient communities project. Centre for Environmental Change and Human Resilience, University of Dundee, Dundee

Folke CT, Olsson P, Norberg J (2005) Adaptive governance of social–ecological systems. Annu Rev Environ Resour 30:441–473

Galicia-Gallardo AP, González-Esquivel CE, Castillo A, Monroy-Sánchez AB, Ceccon E (2019) Organic hybiscus (*Hibiscus sabdariffa*), social capital and sustainability in an indigenous non-governmental organization from La Montaña, Guerrero, Mexico. Agroecol Sustain Food Syst 43(10):1106–1123. https://doi.org/10.1080/21683565.2018.1539694

Gaussens P (2018) The other red mountain: opium poppy cultivation in Guerrero. Text 71:33–69. https://doi.org/10.5154/r.textual.2017.71.003

González-Salazar C, Martínez-Meyer E, López-Santiago G (2014) A hierarchical classification of trophic guilds for North American birds and mammals. Rev Mex Biodivers 85(3):931–941. https://doi.org/10.7550/rmb.38023

Hall B (2001) I wish I were a poem of practices of participatory research. In: Reason P, Bradbury H (eds) Handbook of action research. SAGE, London, pp 171–178

Heindorf C (2011) Analysis of the agrobiodiversity of home gardens in the tropical regions of Mexico. PhD Dissertation, Universidad Autónoma de San Luis Potosí

Hernandez-Muciño D, Sosa-Montes E, Ceccon E (2015) *Leucaena macrophylla*: an ecosystem services provider? Agrofor Syst 89(2):163–174. https://doi.org/10.1007/s10457-014-9751-0

Hernández-Muciño D, Borda-Niño M, Santiago B, Rodríguez R, Rodríguez A, Muciño M, Ceccon E (2018) La Comunidad Me'Phaa Construye su Futuro: Agroecología y Restauración como Herramienta del Desarrollo sustentable. In: Merçon J, Ayala-Orozco B, Rosell JA (eds) Construyendo lo Común desde las Diferencias. Experiencias de Colaboración Transdisciplinaria para la Sustentabilidad. UNAM, CopIt-arXives, Mexico, pp 66–79

INEGI (Instituto Nacional de Estadística, Geografía e Informática) (2007) IX Censo Ejidal. México

INEGI (Instituto Nacional de Geografía y Estadística) (2010) Censo de Vivienda. INEGI, Mexico

INEE (Instituto Nacional para la Evaluación de la Educación) (2005) Porcentaje de población indígena, porcentaje de población hablante de lengua indígena porcentaje de monolingües. https://www.inee.edu.mx/wp-content/uploads/2019/03/CS04-2005.pdf

Lave J, Wenger E (1991) Situated learning: legitimate peripheral participation. Cambridge University Press, Cambridge, UK

López-Ridaura S, Masera O, Astier M (2002) Evaluating the sustainability of complex socio-enviromental systems, the MESMIS framework. Ecol Indic 2:135–148. https://doi.org/10.1016/S1470-160X(02)00043-2

López-Ridaura S, van Ittersum M, Masera O, Leffelaar P, Astier M, van Keulen H (2005) Sustainability evaluation. Applying ecological principles and tools to natural resource management systems. In: Maples AD (ed) Maples sustainable development: new research. NovaScience Publishers, Inc., New York, pp 139–167

McNiff J (2013) Action research. Principles and practice. Routledge, Londres

Melo FPL, Arroyo-Rodríguez V, Fahrig L, Martínez-Ramos M, Tabarelli M (2013) On the hope for biodiversity-friendly tropical landscapes. Trends Ecol Evol 28(8):462–468. https://doi.org/10.1016/j.tree.2013.01.001

Méndez-Toribio M, Martínez-Garza C, Ceccon E, Guariguata MR (2018) La restauración de ecosistemas terrestres en México: situación actual, tendencias, necesidades y oportunidades. Center for International Forestry Research (CIFOR), Documentos Ocasionales 185

Miramontes O, DeSouza O, Hernández D, Ceccon E (2012) Non-Lévy mobility patterns of mexican me'phaa peasants searching for fuel wood. Hum Ecol 40(2):167–174. https://doi.org/10.1007/s10745-012-9465-8

Miranda F, Hernández XE (1963) Los tipos de vegetación de México y su clasificación. Bol Soc Bot Méx 28:29–179

Montagnini F (2006) Homegardens of Mesoamerica: biodiversity, food security, and nutrient management. In: Kumar BM, Nair PKR (eds) Tropical homegardens. Springer, Dordrecht, pp 61–84

Montañez-Escalante P, Ruenes-Morales R, Jiménez-Osornio JJ, Castillo A, Chimal-Chan P, Burgos LL, Cordero WA, Estrada Medina H (2012) Los huertos familiares en Yucatán. In: Vázquez-Dávila MA, Lope-Alzina D (eds) Aves y hueros de México. CONACYT: Red de Etnobiología y Patrimonio Biocultural, Instituto Tecnológico del Valle de Oaxaca, Carteles Editores, Mexico

Muro M, Jeffrey P (2008) A critical review of the theory and application of social learning in participatory natural resource management. J Environ Plan Manag 51:325–344. https://doi.org/10.1080/09640560801977190

Navarro-Siguenza AG (1998). Distribución geográfica y ecológica de la avifauna del estado de Guerrero, México. PhD Dissertation, Ciudad de México: Universidad Nacional Autónoma de México

Navarro-Sigüenza AG, Rebón-Gallardo MF, Gordillo-Martínez A, Peterson AT, Berlanga-García H, Sánchez-González LA (2014) Biodiversidad de aves en México. Rev Mex Biodiv 85:476–495. https://doi.org/10.7550/rmb.41882

Pahl-Wostl C, Hare M (2004) Processes of social learning in integrated resources management. J Community Appl Soc Psychol 14:193–206

Park P (2001) Knowledge and participatory research. In: Reason P, Bradbury H (eds) Handbook of action research. SAGE, London, pp 81–90

Ramos M (2005) La influencia de los aspectos sociales sobre la alteración ambiental y la restauración ecológica. In: Sánchez O, Peters E, Márquez-Huitzil R, Vega E, Portales G, Valdés M, Azuara D (eds) Temas sobre restauración ecológica. Instituto Nacional de Ecología (INE-SEMARNAT), México, pp 31–45

Reason P, Bradbury H (2001) Handbook of action research. London. Sage

Rico-Gray V, Garcia-Franco JG, Chemas A, Puch A, Sima P (1990) Species composition, similarity and structure of Mayan homegardens in Tixpeual and Tixcacaltuyub, Yucatan, Mexico. Econ Bot 44:470–487

Salgado O, Borda-Niño M, Ceccon E (2017) Uso y disponibilidad de leña en la región de La Montaña en el Estado de Guerrero y sus implicaciones en la unidad ambiental. Madera Bosques 3:121–135. https://doi.org/10.21829/myb.2017.2331473

SEDESOL (Secretaria de Desarrollo Social) (2018) Diagnóstico del Programa de Empleo Temporal. https://www.gob.mx/cms/uploads/attachment/file/32298/Diagnostico_PET_1_.pdf

SER (Society of Ecological Restoration) (2004) SER International primer on ecological restoration. Washington, DC. SER

Sierra M (2007) Justicia indígena y Estado: Retos desde la diversidad. In: Robinson CS, Tejera SH, Valladares L (eds) Política, etnicidad e inclusión digital en los albores del milenio. Porrua, Porrua, Mexico

SMN (Servicio Meteorológico Nacional) (2013) Reporte Anual 2013, Coordinación General del Servicio Meteorológico Nacional. http://smn.cna.gob.mx/climatologia/analisis/reporte/Anual2013.pdf

Swantz M, Ndedya E, Saiddy Msaiganah M (2001) Participatory action research in Southern Tanzania, with special reference to women. In: Reason P, Bradbury H (eds) Handbook of action research. SAGE, London, pp 386–395

Tscharntke T, Klein AM, Kruess A, Steffan-Dewenter I, Thies C (2005) Landscape perspectives on agricultural intensification and biodiversity - ecosystem service management. Ecol Lett 8(8):857–874. https://doi.org/10.1111/j.1461-0248.2005.00782.x

Vargas Cárdenas F (2018) El papel de los sistemas agroforestales tradicionales como corredores biológicos para aves: el caso de los traspatios restaurados por la organización no gubernamental indígena Xuajin Me'phaa en el estado de Guerrero. MSc Dissertation, UNAM. Mexico

Wenger E (1998) Communities of practice: learning, meaning, and identity. Cambridge University Press, New York

WRB (2007) World reference base for soil resources 2006, first update 2007. FAO, Rome

Chapter 9
People and Nature Conservation: Participatory Praxis in the Planning and Management of Natural Protected Areas

L. A. Bermejo, J. R. Lobillo, and C. Molina

9.1 Introduction

Participation has become one of the cornerstones of development for over four decades (Chambers 2007). Nowadays, nobody can imagine any development project without participation activities. However, participation encompasses different ideas, concepts, aims, and objectives, and, therefore, its implementation ranges from information providing to self-mobilization and empowerment of civil society. In short, most agencies and institutions have taken and included participation in their activities but using different concepts, definitions, motivations, and aims.

There are several reasons to explain the emergence of participation in most development projects in the last 40 years. However, failures of modernization, the mainstream approach to development, and environmental problems and challenges (Kapoor 2001) are at the core of changes in perceptions about the role of local people in development. As well as these arguments, participation has also arisen in the context of the emancipation of disadvantaged cultures and the recognition of the need for local societies to drive their own development. In this context, participation is not proposed as a novel way to cope with the failure of modernization but as the right of societies to define and implement their own concepts of development. Thus, whereas current approaches to participation are the continuing steps of the

L. A. Bermejo (✉)
Departamento de Ingeniería Agraria, Náutica, Civil y Marítima,
Escuela Politécnica Superior de Ingeniería, Avenida Ángel Guimerá Jorge, s/n,
Universidad de La Laguna (ULL), San Cristóbal de La Laguna (Tenerife), Spain
e-mail: lasensio@ull.edu.es

J. R. Lobillo
Agencia de Servicios Sociales y dependencia de Andalucía, Junta de Andalucía,
Sevilla, Spain

C. Molina
Private consultant, San Cristóbal de La Laguna, Spain

C. Baldauf (ed.), *Participatory Biodiversity Conservation*,
https://doi.org/10.1007/978-3-030-41686-7_9

paradigm of development and modernization (without changing foundations), new concepts of participation, as a right, involve new roles for local societies in development. These new concepts entail redefining local, knowledge, society, and culture hierarchies, as well as relationships with nature and cultures and others. Therefore, these ideas not only modify the implementation of participation but also put into question our relationships with nature and with non-Western societies and cultures.

We propose that participation in natural protected areas and other initiatives of nature protection have not always implied significant changes in the engagement of local societies because the basic definitions and concepts have not changed. To a certain degree, participation works as an ad hoc hypothesis in science, but it does not entail a new paradigm. Moreover, we question if Western society is ready to establish a non-hierarchical dialogue with cultures that coexist or have coexisted with nature and if this current participation enthusiasm (as an ad hoc tool is just to maintain the current paradigm in order to cope with failures of the Western model) could be an obstacle instead of driving the development of participation as a right.

In this chapter, we discuss the role of local culture and natural protected areas in the mainstream modernization process of development. In this context, natural protected areas represent an inextricable part of modernization, and therefore they are constructed (planned and managed) using the same foundations. After that, we discuss how different ways of participation are implemented according to different definitions and concepts of knowledge, local, nature, and participation itself. The way public participation is applied responds to the requirements and interactions of different approaches that include and frame social, political, economic, cultural, and psychological elements among others but also introduce sources of bias (sometimes unconscious or unknown). We propose that these approaches guide and explain the whole implementation of participation in the planning and management of natural protected areas. Moreover, we describe participation as a means to achieve external goals as well as participation as being a right and as an aim in itself. Both are studied as ideal categories in three different analytical approaches. We study several practical cases in order to depict the results of the implementation of different concepts of participation in several development projects in natural protected areas.

9.2 Theoretical Framework

9.2.1 Modernization and the Enforcement of a Western Model

Modernization, as a theory about the evolution of society, entails a technological change from "traditional" and lagged stages to "modern" and advanced stages. This idea is mainly based on two key issues. On the one hand, the social construction of the "traditional," "local," and "ethnical" as lagged structures must be changed toward Western-like society as the only feasible model (Contreras 1984; Matijasevic and Ruiz 2013). Therefore, modernization implies the modification of the original local and traditional society through absorption, transformation, or any other way

(Sánchez de Puerta 1996 pag. 248, Rogers and Svenning 1969). On the other hand, such a technological change is based on science, which is considered the most powerful tool for development according to the modernization concept (Hardeman and Jochemsen 2012). This means that other sources of knowledge are not sufficient for development, among other reasons, because other knowledge is constructed in backward contexts. Thus, from this point of view, local (traditional) societies are not able to produce feasible tools for their development and evolution, whereas science and technology can provide these tools. Moreover, science and scientific structures, institutions, and agencies have become powerful mechanisms to legitimize the imposition of certain transformation processes over so-called lagged and unenlightened societies. However it should be noted that social injustice begins with cognitive injustice resulting from the enforcement of one knowledge over another (Sousa 2011) giving rise to a knowledge hierarchy in which some societies are placed below "scientific" ones.

Rural societies are included in the underdeveloped society group from the point of view of modernization. Therefore, peasants would be described as closed, unwilling to innovate, unproductive, and inefficient (even production data have been modified in order to justify technological transformation (Van der Ploeg 1990)), and sometimes pejorative judgments have been published (Contreras 1984), probably as a justification for the transformation of rural society (Sánchez de Puerta 1996). Indeed, modernization is still driving most of the current initiatives for rural development. In this context, success is usually assessed through the levels of intensification, specialization, scale (Hardeman and Jochemsen 2012), and commodification of farms, and explanatory factors such as illiteracy are widely considered as decisive (Pingali 2012; Feder and Savastano 2017).[1] Moreover, local society is perceived as nature destroyers instead of custodians (Fairhead et al. 2012), which reinforces the need for transformation.

This construction of ideas, regarding "lagged peasants," has supported transformation processes and the spread of homogeneous Western society. In the Lévi-Strauss sense, modernization is a deculturation process through universalization that destroys and displaces particularisms. Even recent coexistence assumptions are ethnocentric, since Western dominant culture has been unable to construct non-hierarchical relationships with other societies. Still, in certain contexts, traditional production systems are considered and studied as cultural resources that have to be protected and developed. Sometimes, this concept reinforces the perception of these systems as non-modern styles of agriculture and therefore as lagged ones. This view is also ethnocentric and consolidates current hierarchies, because it maintains the dichotomy old (ancient) vs. modern that supports cultural transformations.

[1] It seems a circular argument, since response and explanatory variables are previously related and linked. Both are embedded in the same paradigm, since commodification requires technology (non-local technology) that can only be comprehended through Western semantics and meanings. These semantics establish the border between illiterates and literates and do not allow illiterates to be considered literates in other semantic domains, namely, the "local." In this context, the ecology of wisdom proposed by Sousa (2011) has no place.

Science, as the only legitimate source of knowledge about the modernization paradigm, supports these universalization processes thanks to generalization and unbiased principles. However, both these principles are widely disputed (Ravetz 2006; Henríquez 2013). Generalization assumes that scientific findings can be successfully implemented regardless of previous conditions or that these conditions can be controlled and modified in order to ensure success (the world in a laboratory (Ravetz 1999)). Success is achieved when local (and lagged) societies become modern societies through removing (or modifying) all (or most) of the characteristics that anchored them to the past. However, uncertainties (technical, methodological, and ethical) grow as the scale of intervention (more properly its extension) increases, since emergent (and unknown) factors arise (Gibson et al. 2000; Ruiz Rivera and Galicia 2016) and results differ from expectations and predictions. Scale, as extension, defines the spatial and temporal dimension of intervention (and decision-making scope) (Ruiz Rivera and Galicia 2016). Increasing dimensions implies the emergence of factors that modify the system's observable behavior, and this behavior becomes (or seems) more random as the intervention extension increases because uncertainties increase as well.

However, unbiased science is a controversial issue. Biased assumptions question one of the bases of research procedures, namely, the decoupling between observation and interpretation (it implies that observation as unbiased phenomenon drives interpretation). Decoupling supports the idea that observation is an unbiased process of a physiological nature that is not affected by the observer's culture, ideas, or perceptions (Henríquez 2013). However, other research points out the existence of observational language, since observation is inextricably linked to interpretation. This link implies that observation itself is biased by research pre-assumptions that condition not only what part of reality is observed (and what is not) or what is taken into account (and what is not) and also how it is observed (even measurement instruments assume some theories that aim to prove something (Echeverría 1999)). Therefore, the influence of nonscientific factors on observation-interpretation (Henríquez 2013) is an important part of decisions. The importance of non-rational factors also increases as scale (extension) increases in complex systems. Therefore, on large scales in complex systems, decision-making based on science presents problems of generalization and significant perception and bias effects. Ravetz (2006) proposes that decision-making processes face cumulative uncertainties (from technical to ethical uncertainties) as what is at stake increases (Fig. 9.2). Thus, science, by itself, is unable to provide solutions on large scales, probably because of the complexity and number of emergent factors. Indeed, when limitations of science in decision-making are not well established (and it maintains its relevant role at large scales), bias and specific interests arise with regard to other types of knowledge, as Henríquez (2013) suggests.

Ecosystems and societies are complex systems because their structures and dynamics are driven by a huge number of variables and interactions. Thus decisions reflect the consequences of the limitations and constraints of science, which compromise their effectiveness and usefulness.

The modernization paradigm proposes unique, general, and unbiased development models for a huge range of different regions and conditions. These models are mainly characterized by external modification of local societies and cultures and whole-scale replacement where transformations are not possible. Intensification, scale, specialization (scale economies), and commodification of all production factors are the main way of implementing a modernization paradigm, since this maximizes efficiency of resource allocation and usage (human, financial, and natural resources among others), such as in industrial processes and structures and even in agriculture (Van der Ploeg 2014). These are the main principles that drive most of the decision-making processes in different domains, such as nature conservation, rural development, and land planning and management.

However, these principles and models may have been successful for certain regions, but they have not been successful for others mainly due to two reasons. First, modernization requires preconditions for successful implementation such as resource availability and investment, and, second, unlike industrial processes, nature conservation, rural development, and land planning and management are complex systems, and, therefore, decision-making is affected by uncertainties and ignorance that science is not able to totally control (Ravetz 2006). Therefore, as many authors have showed, the modernization paradigm has failed in many places and situations (Van der Ploeg 1990; Sánchez de Puerta 1996; Sevilla 2013), not only in the socio-economic domains but also in environmental domains (Martinez-Alier et al. 2016), as is widely known.

9.2.2 Natural Protected Areas as Implementation of Modernization Models

The implementation of natural protected areas is the practical embodiment of a Western and ethnocentric perception of nature (Cronon 1996; West et al. 2006). This perception is profoundly rooted in modernization principles and models. In this way, tensions and strains between conservationists and non-conservationists are Western society conflicts where local communities are part of the problem in dispute. However, beyond these conflicts, most traditional societies protect natural resources against governments and companies, since nature (and local systems of appropriation) supports its own social reproduction (Martinez-Alier et al. 2016). Moreover, although initiatives have been proposed by local and native societies (as in many natural protected areas), planning and management are based on Western perceptions about nature and on the prominent role of science. Therefore, though claims by local communities are sometimes accepted, management often tends to be enforced from the outside. Does this therefore entail a softer kind of dispossession from the point of view of local societies? (Sevilla 2014).

The modernization principles mentioned above are present in natural protected areas across the world. Land use planning and management and their regulations are tools for the replacement and dispossession of local societies to ensure conservation of natural resources (West et al. 2006). In fact, Delgado et al. (2014) propose that green grabbing is one of the four types of land grabbing that, as other grabbing types of modernization, can cause local culture displacement and inequalities (Fairhead et al. 2012). Nowadays, decisions are taken on a supra-local level, and local and traditional decision structures have no role in the modern age (Sevilla 2014). Replacement and dispossession are justified because Western society is perceived as being at the uppermost level of evolution in comparison with local cultures, among other reasons because Western society is considered to have developed an unbiased, replicable, and generalizable tool such as science. From this point of view of modernization, only science and scientific structures provide proper and suitable solutions for development.

Although dispossession in a wide sense is one of the traits of modernization, another characteristic of natural protected areas, as modern institutions, is the specialization of procedures of planning and management (in the same way as other specializations). Natural protected areas are used for (specialized in) nature conservation as a repair strategy to compensate for the damage inflicted by economic growth, as Fairhead et al. (2012) suggest. These concepts are embodied in land-sparing structures (Grau et al. 2013), where space is designed as a mosaic of specialized uses. Some areas are protected and isolated from human activities, whereas others are used for intensive and high-yield activities (Kremen 2015) at the expense of environmental impact. This conception of land distribution fits with the modern concept of nature as opposite to non-natural (industrial-urban society), where territories are specialized and grabbed for different purposes (Delgado et al. 2014). Nature conservation is one of these purposes (Fairhead et al. 2012). Land-sparing structures are far from nature-human integration of traditional and peasant economies (Van der Ploeg 2015 pag. 71; Toledo 1990), which are closer to land-sharing approaches where strategies for establishing uses that conserve resources and ecosystems (overlapping different land uses) are proposed (Bermejo and Lauenroth 2012; Edwards et al. 2014) with differing results. Current natural protected areas are unalienable parts of modernization that are needed to defend nature.

Indeed, natural protected areas shape the immunological protection against the effects of modernization (and humans) on nature and, therefore, they are part of modernization itself (Han 2016) (even ecological monitoring is a medical metaphor). Threats (modernization-industrialization and humans) and protection tools (natural protected areas) for victims (nature and local population if this is ancient enough to seem part of nature) share most of the principles and structures, namely, a unique model based on science that entails local culture dispossession and conservation as specialized land use. Both are profoundly ethnocentric, and they minimize the role of local societies in decision-making. Even some tributes and homages contribute to this aim in a long farewell ceremony that arose alongside modernization (Bermejo and Cubas 2019).

These assumptions are reinforced by the idea (and scientific theories) of the equilibrium of nature as a static stage. Ecosystems tend toward this stage, and native species coexist together, and biodiversity reaches its maximum point, and nature becomes wild and pristine again. See theories such as competitive equilibrium (Hardin 1960), island biogeography (Macarthur and Wilson 1963), the productivity-stability hypothesis (Connell and Orias 1964), the stability-time hypothesis (Sanders and Hessler 1969), among others.[2] In this way, Huston (1994) stated, "the concept of equilibrium was also consistent with a philosophical inclination among ecologists to believe that the continued existence of natural communities reflected some sort of balance of nature, stability and equilibrium." According to these theories, humans, as disturbances, break this equilibrium and, therefore, threaten nature and biodiversity. These models shape mainstream conservation that meets with the Western idea of a wilderness with no humans[3] (Cronon 1996; West et al. 2006) and are implemented and established as natural protected areas specialized in equilibrium preservation.

In this context, most land use planning for nature conservation is established along a negative gradient between conservation and human usage (the case of natural protected areas in Canary Island, Spain, in Fig. 9.1). In this framework, conservation levels are defined as the distance from some kind of static equilibrium stage (harmonious platonic stage), where functions and structures of ecosystems and resources are protected and maintained (Huston 2014). Therefore, ecosystems range from disturbed stage to natural stage, where biodiversity is highest. Human usage is directly related to disturbance and transformation of natural conditions and therefore to equilibrium modification. Human uses range from non-use to industrial-residential use (constructions and others), where disturbance is highest and ecosystems are degraded and biodiversity is lowest. In many places, natural protected areas are classified according to the level of nature protection and/or the level of prevention against human use, ranging from residential use to non-use areas (in the Canary Islands, natural protected areas range from rural parks to integral nature reserves (Gobierno de Canarias 2017)) (Fig. 9.1). Differences among protected areas are related to the proportion of the types of zones within each area. In the Canary Islands, these zones range from special use zones, where construction is permitted, to exclusion zones, where access is not permitted (Gobierno de Canarias 2017)) (Fig. 9.1).

Ultimately, use and conservation are mutually exclusive in natural protected areas like land-sparing structures. Humans lose presence gradually along this gradient and have no role in nature only as aseptic observers (another medical metaphor) and become the audience who attends the spectacle of nature.

Therefore, human use as an element in the management and conservation of ecosystems and biodiversity is not feasible in the context of this conceptual universe that emanates from the conception of nature as an untouchable paradise (Cronon 1996). Thus, natural protected areas are the result of an equilibrium model implementation. According to methodological pluralism theories (Kuhn 1970; Lakatos

---

[2] Here we can question whether perceptions and ideologies shape customized theories or unbiased theories and if their results match (or shape) perceptions and ideologies and then with culture.

[3] Wilderness is the last place that humans have not destroyed or transformed and, therefore, the last haven of pristine nature where stressful, busy citizens can escape from modern, crowded, and unhealthy cities, as Cronon (1995) proposed.

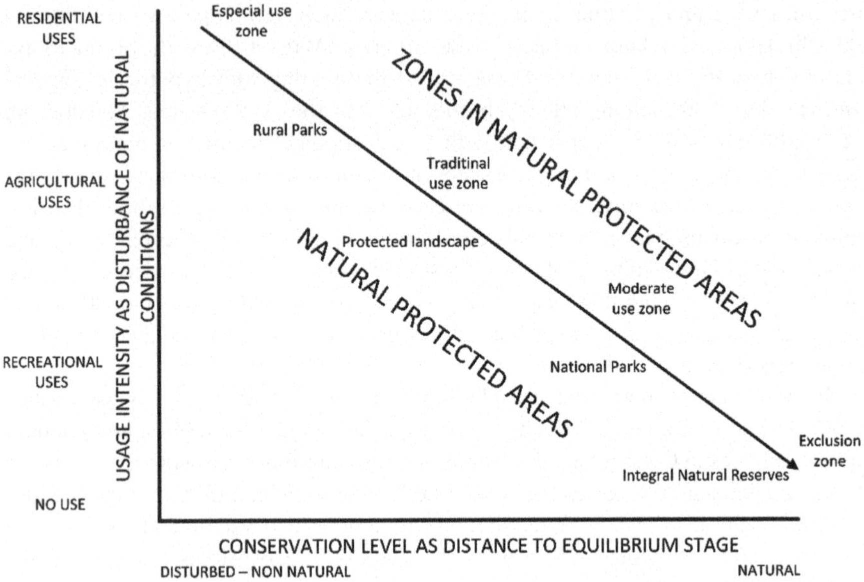

Fig. 9.1 Relationship between protection level and human use level in the Canary Islands Natural Protected Areas Act (Gobierno de Canarias 2017)

1970; Feyerabend 2010), social and cultural factors condition scientific objectives and procedures in order to support them, their changes, and replacements. Finally, nature as a cultural, social, and historical construction can drive scientific concepts, methods, findings, and implementation.

9.2.3 Participation and the Role of People in Decision-Making

Modernization as a planned transformation of society toward a general and unique model entails the removal of the role of local societies in decision-making. First, locals do not provide useful information for modernization, since science is the only legitimate source of knowledge. Second, the Western and modern model is desired worldwide. However, participation has been gaining relevance in different domains since the 1980s because of the contributions of different authors from different disciplines. There are mainly two sources of change and a redefinition of the role of local cultures in social evolution. One source gathers all of those ideas related to the social justice of oppressed classes, related to doubts about usefulness of scientific research for people's emancipation (Lewin 1946; Tax 1960; Fals-Borda and Rahman 1991) and related to the definition of underdevelopment as a psychosocial issue and local communities' inability to change their surroundings (Illich 1969 cit. Oakley 1991, pag. 26). In this context, participation acquires the meaning of a political right (beyond the right to be informed) related to empowerment (Cleaver 1999) and the relevant roles in decision-making and as way for development.

Another source of change in the concept of participation comes from the failure of development policies in Latin America and Africa as result of the implementation of the modernization model (Chambers 1983; Bunch 1997). Here, participation is more focused on efficiency and effectiveness (as a tool for achieving better outcomes, conflict control, and others) (Cleaver 1999) and of external objectives as a way of development. These two concepts are extremes on a gradient of participation from a political right to a means to enhance efficiency for modernization (Chambers 1994) rather than closed and ideal categories.

There are three interesting approaches that involve a different classification of participation ideas that place typologies of intervention within different theoretical frameworks. Arnstein (1969) established a gradient of citizen participation in her seminal paper, from non-participation to citizen control in decision-making. In practical terms, she described the change from manipulation as a non-participative process and the right to be informed (in a passive role) as tokenism to achieving a relevant role of citizens in the whole process of decision-making. She proposed the well-known eight-rung ladder of citizen participation. The second approach discriminates participation as a means to achieve external aims from participation as an aim that empowers local people in a long-term process (Oakley 1991).

Different categories and types do not exclude each other, and they are well integrated in the third approach. The framework of post-normal science places participation within the relationship between high stake decisions (scale) and uncertainties that are faced in the decision-making process (Fig. 9.2) (Ravetz 1999). Participation

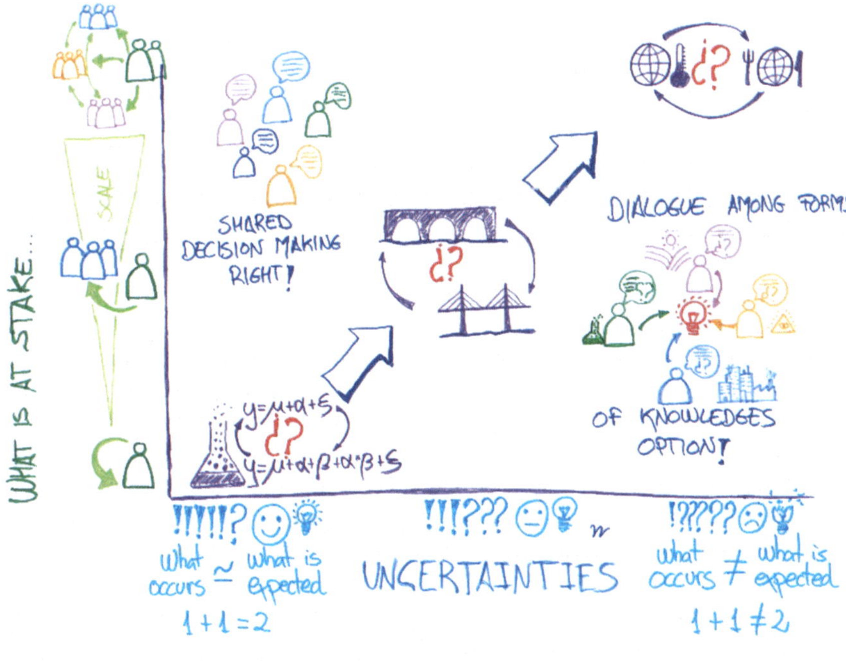

Fig. 9.2 Relation between what is at stake and system uncertainties

is a right when what is at stake is high, even though uncertainties are low and science provides feasible and effective solutions (Ravetz 2006). In this context, decision-making must be shared and agreed by different groups. In addition, when we face high levels of uncertainties and what occurs is unpredictable (because of its complexity), participation is not just a right but it is an option for decision-making through dialogue between forms of knowledge and perceptions. The priority between climate change and hunger (complex matters and high decision stakes) is not only an issue for science, but also these decisions involve everybody, because people's participation is a citizen right and because nonscientific knowledge could provide proposals as a way to cope with such complexity. In this context, participation as a right is extended with its function in knowledge providing.

The key question behind these views, ideas, and perceptions about the relevance of participation is: why do people have to participate in development processes? The answer involves not only the procedures applied but also the desired outcomes of any decision-making process and, ultimately, it involves the power relationships between stakeholders and their roles. Therefore, concept definition and understanding are the key elements of participation.

We need to analyze two concepts about participation that combine the ladder proposed by Arnstein (1969), classification by objective proposed by Oakley (1991), and post-normal science approaches by Ravetz (1999). On the one hand, on the lower rung of Arstein's ladder is the more utilitarian view of participation that fits with the participation as described in Oakley's classification, since intervention efficiency and effectiveness are the main aims. On the other hand, participation can be conceived as an aim in itself, also proposed by Oakley, and is close to the concepts included on the higher rungs of Arstein's ladder, where citizens are subjects of development instead just objects. Post-normal science includes both approaches and assigns them different roles.

Participation as a Means: The Utilitarian View of Local Participation for Efficiency

The utilitarian view of participation encompasses two ideas. First, participation is a means to achieve aims of the external projects, policies, and initiatives (in order to improve project and policy efficiency). Second, since goals are predefined, people become the objects of development (as more or less passive beneficiaries) rather than active subjects (Oakley 1991). In addition, participation contributes to efficiency in two aspects related to knowledge. On the one hand, native knowledge can be used to cope with the complexities of local conditions (as result of limitations of the generalization capacity of science), not only in terms of information quality but also in terms of time (rapid rural appraisal and other methods improve time-reliability efficiency (Chambers 1992)). This is participation by providing knowledge (information). On the other hand, the capacity to respond to local needs (at least in part) and to involve people enhances the likelihood of the project, initiative, or policy's success. This is participation by local support and contribution. The aims are related to technology adoption (Sánchez de Puerta 1996), public investment and facilities, conflict control in land and resource planning (for instance, natural protected areas) (West et al. 2006; Sevilla 2014), engagement in nature management, and other activities such as attendance at training activities.

According to the seminal work of Arnstein (1969), participation as local support and contribution could be considered within a non-participation framework (manipulation and therapy) and tokenism[4] (consultation, informing, and placation), since final decisions are taken by outside power holders. Power is held through governance rules and, mainly, through hegemony (moral and intellectual guidance) (Gramsci (2014) and Foucault (2008) cit. Sevilla 2014). Although participation level could be high within this framework, objectives, problems, and some guidelines and general strategies are predefined by power holders (and their concepts of modernization, nature, and science). Indeed, manipulation and therapy are explicit mechanisms of imposition that aim to place the local population on a modernization path, any consultation, informing, and placation only aim to achieve these objectives through the appearance of participation instead of explicit imposition. In many cases, conflict control in natural protected areas is the main goal of participation activities that do not go beyond consultation or informing. In the end, the appearance of participation and power is part of the process of governance in the sense of Foucault, and rules are peacefully introduced in society and culture in order to avoid conflicts before they occur (Sevilla 2014). This concept of participation is adopted by most development projects and citizens' participation in policies and decision-making such as the planning and management of natural protected areas.

Participation as an Aim, Participation as a Right

The highest rungs on the participation ladder proposed by Arnstein (1969) are related to local people's power and the right to play a relevant role in decision-making. Therefore, participation becomes an aim in itself and a means to achieve population empowerment instead of a means to achieve external objectives (Oakley 1991). In this context, people take the role of subjects of development instead of objects in projects. Development begins to frame and to include concepts such as identity, political participation, change capacity (Terluin 2003), cognitive justice (Sousa 2011), and autonomy (Van der Ploeg 2015). Inclusion of these factors contributes to development by overcoming Illich's (1969 cit. Oakley 1991, pag. 26) conception of underdevelopment as self-perception of irrelevance. Participation as an aim is related to citizen control, delegated control, and partnership (Arnstein 1969) that is established as a right when what is at stake is high (even though uncertainties are low) (Ravetz 2006) (Fig. 9.2).

[4] Tokenism is defined as actions that are the result of pretending to give advantage to those groups in society who are often treated unfairly, in order to give the appearance of fairness.

However, real participation as a right requires leaving aside Western hegemony as a way of development and ensuring the inclusion of other knowledge, ideas, and perceptions of "nonexisting cultures" according to the sociology of absences proposed by Sousa (2011). Participation as a right entails new processes of production of knowledge and new relationships among different ways of knowing. This idea drives a new creative dialogue throughout the ecology of knowledge and intercultural translation (Sousa 2011) based on participation as an aim in itself and, therefore, as a right. Going further, participation is a way to evolve toward a fairer world in inter- and intra-generational terms. Therefore, it is not related to replacing the old hegemony with a new one but participation is related to the dialogue among cultures on equal and non-hierarchical terms in order to cope with complexity and uncertainties (not only technical and methodological but also ethical and epistemological ones) (Ravetz 2006).

Therefore, participation requires a collective definition of basic concepts and terms and even the problem to be solved. In the case of natural protected areas, not only concepts such as conservation, resources, impact, disturbance, and others have to be jointly defined but also development and quality of life (because they are defined in the context of Western culture, its instruments, and its knowledge hierarchies). Without this creative dialogue, participation will not transcend the participation simply as a means or as tokenism.

9.3 Current Status of Participation in Development Practices in the European Union

Participation has become an indisputable part of policies and development programs around the world. Nowadays, there is no intervention without participation activities, in many cases as a legal requirement (Aarhus Convention) (The European Parliament and the Council of the European Union 2000, 2006). In other cases, it is part of methodologies for rural development programs (Agricultural Policy of the EU (Navarro et al. 2014)).

Despite attempts to enhance and improve public participation, these initiatives have not overcome the hierarchical dichotomy between government and people. This also entails other dichotomies such as science vs. local knowledge or global vs. local identities among others. These dichotomies remain because decisions are governments' responsibility and, therefore, taking into account participation results is a technical-political decision (Valencia 2009) that means the decisions are not jointly taken but are considered (and accepted or rejected) by power holders (government, technical staff, science, and other actors of higher levels of the decision-taking hierarchy). Although, according to law (Aarhus Convention), acceptance or rejection of public contributions have to be justified and argued, the cognitive hierarchy (science vs. other knowledge) underpins these decisions, since legitimacy of the different sources of knowledge to accept or reject is not the same (not even similar) and one exerts power over others.

Therefore, participation is not only a matter of including a new element in the common concepts of development, environment, or quality of life (using participation as new adjective for ancient nouns as Sousa (2011) proposes, such as participative development, participative planning, participative budget, and others). Participation is a matter of transformation of the role of local and people (their way of being, thinking and feeling, their knowledge, their perceptions of time and space, etc.) in society's evolution (and include other nouns as *umma*, *dharma*, *sumac kawsay* for human rights, dignity, quality of life, etc., instead of re-qualifying old terms) (Sousa 2011). Therefore, participation under its mainstream concept does not overcome the role of tokenism nor does it achieve any degree of citizen and public power (Arnstein 1969).

In Spain, natural protected areas like any other plan, program, or policy require a process of Strategic Environmental Assessment (SEA) to be approved and implemented (The European Parliament and of the Council of the European Union 2001). As mentioned, participation is one of the cornerstones of the SEA Act, but the public's role is limited to being consulted and informed about the plan once most of the decisions (and the most important decisions) have already been taken. Consultation and information are on the lowest rungs of the participation ladder and keep people as objects of initiatives. We could place consultation and information as different degrees of tokenism according to Arnstein's (1969) ladder, since sometimes one of the objectives is the control of social conflict, mainly in natural protected areas (West et al. 2006). However, participation is a right included in planning and policy-making procedures. Transposition of the SEA Act into Spanish national legislation specifies not only the right to be informed and consulted but also the usefulness of local knowledge and ideas. However, these statements are not able to go beyond the utilitarian view of local knowledge for effective improvement of programs, plans, and policies. Finally, the novel role of local wisdom in legal procedures is not related to non-hierarchical dialogue among different types of knowledge (Fig. 9.2), rather it is information used for external aims and decision-making processes. Although local people's role has become more relevant in recent decades, citizens remain as objects instead of subjects of policies, since they do not have a real and transformative influence on decision-making, not even when what is at stake is high in social, economic, and environmental terms (Fig. 9.2) such as in natural protected area implementation.

9.4 Participation in Practice: From the Desirable to the Feasible

Within this framework, practical experiences of participation can be described using comparative analyses. This allows us to determine internal and external constraints to achieving higher degrees of public power within the concept of participation being considered a right.

We analyze three different cases of participatory processes through a qualitative approach. The comparative analysis consists of (1) a *brief description* of each case including general objectives, funding, and other aspects and (2) a *qualitative approach* to participation processes. The qualitative approach includes a) participation in project stage (from the initiative to the implementation and continuity) and b) contribution to aspects such as self-knowledge, empowerment, and changes in surrounding, among others. The three cases are i) infrastructure and rural development in a natural protected area, ii) grazing management plan in natural protected areas, and iii) strategic plan for a biosphere reserve. All cases are located in Canary Islands, Spain.

9.4.1 Infrastructure and Rural Development in Natural Protected Areas

A local government initiative for an infrastructure project in a specific natural protected area aimed at nature conservation and rural development. These areas are called "Rural Parks." The main objective of the project was the improvement of agrarian structures for rural development, whereas the aim of participation activities was to increase efficiency and efficacy of investments through achieving the best adequacy to local needs and demands. The participants were the local population (and its formal associative structures) and technical staff of the Rural Park. The outcome was a document of prioritized infrastructures, which, however, was not binding.

Participation processes were emphasized in the appraisal stage, though the overall objectives had previously been defined and established. Indeed, the main aim was the modernization through capital investment in a non-local perception of development. Moreover, there was no participation in the implementation or continuity stages, not even in terms of contribution, shared responsibility, or participative monitoring.

In this context, the predefinition of objectives restricted the appraisal process to aspects related to infrastructures, and other issues were excluded. This fact had two main consequences. On the one hand, there was scarce contribution from self-knowledge as a source of empowerment and organization. And, on the other hand, there was fragmented knowledge about the local reality by technical staff and government managers. Therefore, the opportunity to construct an overall view was lost.

During the appraisal, participation was more intensive. Although knowledge and information were biased toward the project's main issue, broader participation would have allowed local views to be captured about the issues. It was a thorough appraisal but too fragmented. The direct consequence was an inability to establish some kind of balance between common and individual benefits by the local people. Many participants attended working groups in order to defend their own interests instead of common benefits. Social construction of fragmented knowledge and its consequences also hindered the local empowerment and organizational capacity. Moreover, there was not a well-organized process of information feedback to local people, which would have increased the general perception of the appraisal as knowledge collecting (extraction) instead of knowledge sharing. However, in some ways, this participation process did contribute to enhancing the efficiency of public investment allocation since most of the proposed infrastructures responded to local demands.

During the implementation stage, priorities established in the appraisal were not totally taken into account. Local people (even technical staff) perceived that their role in development was weak and irrelevant, and, therefore, their local capacity to change (and improve) the reality was questioned widely. Many of the participants in the first implementation phase gave up on the participation activities and stated their disappointment. This fact harmed any empowerment and organizational initiatives.

9.4.2 Grazing Management Plan in Natural Protected Areas

Goat grazing has traditionally led to widespread conflicts in the Canary Islands. These conflicts are particularly relevant in protected areas, since they are mainly assigned for nature conservation, and, therefore, human activities are regarded as disturbances that affect the natural equilibrium. In this context, the social construction view of farmers is mainly negative, and they are often included among threats to nature and biodiversity. Grazing plans were therefore aimed at establishing the characteristics of sustainable grazing management in protected areas. Farmers, different technical staff, and regional and local politicians were engaged in the planning processes. The main outcome was a non-binding document, although grazing areas were recognized in land regulations in force. From the point of view of participation, the main objective was to create a permanent structure for grazing management as a means to increase local people's power in decision-making.

Farmers demanded a grazing management plan because of the pressure of environmental departments to eliminate or limit grazing activity and because of the negative perceptions about farmers. Indeed, grazing as a negative disturbance is included in environmental education campaigns. Therefore, the initiative was from goat grazing farmers. Local government supported and funded the whole project.

Since this grazing management plan was a farmer's initiative, the main problems were locally established and prioritized during the appraisal stage as a result of dialogue and reflection among participant groups, where local knowledge played a relevant role. This contributed to an in-depth, collective knowledge about the local reality, and provided a solid and shared global view of the farming sector, among other reasons because some stakeholders were involved for the first time (to provide other views), and because there was enough time to go deeper in the appraisal. Once problems, difficulties, and constraints were defined and prioritized according to local criteria, activities were designed in the context of larger working groups, where farmers, technical staff, and politicians gathered for planning. Going a step forward, responsibilities for each single activity were assigned to different participant groups that committed to implementing or working toward implementing them. Assigning knowledge and responsibilities contributed to local empowerment and organizational capacity and, therefore, to local perceptions of change capacity as well. Moreover, participants were conscious of the real difficulties to achieve certain objectives and of the real capacity for implementation or to influence implementation. Therefore, some activities were excluded because of their difficulty, whereas others were undertaken as feasible. This was possible because participants acquired an overall view of the social, economic, and political environment through group discussion and dialogue among stakeholders.

However, these commitments were not binding for stakeholders. This lack of binding commitments affected the implementation, which did not meet expectations, mainly because priorities were not respected at all. Moreover, appraisal activities led to a weak but hopeful perception of effective local participation in development. This perception of local empowerment was then profoundly questioned because power holders (mainly local and regional political representatives and technical) did not fulfill the agreed commitments in the implementation stage. This fact undermined the initial impetus of the participatory appraisal to consolidate a permanent structure for local participation. Finally, the whole project did not contribute to empowerment and organizational capacity or, therefore, to the capacity to change the reality and environment. To some degree, expectations raised in the appraisal were frustrated in the implementation phase, which negatively affected any chance of changing local roles in decision-making toward a greater capacity and influence on development.

9.4.3 Strategic Plan for La Gomera Biosphere Reserve

Biosphere reserves are a well-known program of UNESCO for nature conservation and development of specific regions through establishing a scientific basis for the improvement of relationships between people and their environments. La Gomera Biosphere Reserve was a local government initiative that aimed to protect the biggest laurel forest in the world (protected as national park since 1981) and ensure its economic development. Although it was a local government initiative, organizations of local society rapidly became involved in the process. However, not only the initiative but also strategic activities were defined and designed by a technical board, as well as performing an analysis of the problems. Therefore, participation was implemented in order to improve the efficiency and efficacy of La Gomera Biosphere Reserve and to ensure the success of the initiative as a predesigned project.

In this context, participation was a way to achieve external goals through informing, consultation and placation (degrees of tokenism). The aims of participation were to include local knowledge and views but only in a small part of the planning phase, since the important and relevant decisions had been made by local government institutions. This fact caused two main effects. First, the lack of participation during the appraisal process did not allow participants an overview of the entire project and of its contribution to local reality. The main consequence was the scarce local involvement in project implementation. Second, the project did not significantly contribute to the empowerment and organization of people, probably because local people's role was restricted to provide information and consultation. Information was not fed back in a formal process of creative dialogue, which meant information was taken but it was not shared. This fact reinforced the role of people as just information providers. In this context, people were objects to achieve the project's goals instead of subjects of development.

Participation was restricted to defining the main strategies based on predesigned activities applying logical framework approach. Activities were prioritized according to the importance and relevance for local development and according to local capacity to implement them (approachable, influential capacity, and out-of-reach activities and strategies). The process makes us aware of the real capacity of influence on the surroundings, and it encouraged people to focus on the approachable strategies. Moreover, participants had experienced this kind of process, and they were familiar with methods and tools, which contributed to the quality of results.

One limitation was the short time to develop the whole process that did not allow a profound and creative dialogue among participants to be established. Therefore, participation was used as rapid tool of information collecting similar to other methods such as the Rapid Rural Appraisal, which does not aim to improve people's empowerment but only aims to obtain information from local people.

Finally, this process did not lead to significant negative effects on people's empowerment and organization, since there were no expectations about the project, partly because the initiative was not totally related to people's needs. On the other hand, project goals were successfully achieved, and a strategic plan was developed.

9.5 Conclusions

As we highlight in this chapter, participation in natural protected areas is determined by the ideas and concept that drives its implementation. It is defined by views about nature (pristine wilderness with no-human vs. human presence in nature), conservation (protection against most human uses to achieve the pristine equilibrium vs. dynamic equilibrium and disturbance), knowledge (legitimacy of science with regard to other knowledge, wisdom, and cognitive justice), and power for decision-making (local rules are replaced by supra-local regulations).

However, regardless of the range of views, participation faces several internal and external factors. Internal factors are dependent on technical staff and local participants. They can be managed and improved, at least within each conceptual framework. Main internal factors are as follows:

1. Tradition or culture of participation. In the Biosphere Reserve case, a tradition of participation made the process easier. However, a lack of culture of participation was a constraint in the processes in the other two cases, not only because of the low expertise in tools and methods but also because of scarce awareness about the meaning of participation. It should be noted that participation is a process in which groups and individuals learn by doing, and each experience is a new step toward better implementation.

2. Procedures. The use of suitable procedures, tools, and methods of participation drive not only the information produced but also contribute to self-knowledge and, therefore, to empowerment and organization. In the end, the right procedures will contribute to a cultural change with regard to the role of people in decision-making. In the case of the grazing management plan in a natural protected area, a wide approach allowed a high level of self-knowledge, and an overview of the livestock sector and surroundings was achieved. However, in the case of rural development in natural protected areas, procedures were focused on infrastructures, and the approach was much poorer. Its contribution to empowerment and organizational capacity was therefore rather limited.
3. Needs. Participatory processes have to take into account local needs in the different stages or phases. Whereas the grazing management plan responded to sector needs, in the case of rural development in natural protected areas and the Biosphere Reserve plan, the main goals were defined previously. However, specific actions responded to local needs. Hence, the first case contributed more to engagement of local people in comparison with the other two cases. All responded to people's needs but at different levels (overall project vs. activities within a predesigned project), which led to different levels of involvement.
4. Expectations. Although expectations can launch a process successfully, they may be negative if they are not met. People gave up on the rural development project, because expectations were not met. However, no expectations discourage people's involvement. Thus, participants have to have an accurate idea of expectations in terms of likelihood of occurrence, scope, and local capacity of influence (Does compliance with expectations depend on people?) such as in the grazing management plan and Biosphere Reserve plan. Therefore, the management of expectations is a key factor in participation.

External factors are characterized by the difficulties and constraints to be managed, because they are related mainly to changes in the power hierarchy. So, they are outside local control. Transferring power is a political issue, and it triggers resistance behaviors throughout the different strategies, yet changes in power structure are at the core of the transition toward real participation. Despite efforts to reach higher levels of empowerment of local stakeholders (empowerment in terms of people power), we do not achieve this aim, mainly because of the obstacles caused by external factors.

Participation implementation is the result of the relation among ideas and the influence of internal and external factors. The key question is what kind of participation are we implementing? And how far do we want (or can) to go with participation?

References

Arnstein SR (1969) A ladder of citizen participation. J Am Inst Plann 35:216–224. https://doi.org/10.1080/01944366908977225

Bermejo LA, Cubas F (2019) El pastoreo ante la expansión urbana, la protección ambiental y la escasez de territorio. Cuad Investig Urbanística 123:24–34

Bermejo LA, Lauenroth WK (2012) Conservation grazing management: a novel approach to livestock management and biodiversity conservation on the Canary Islands. J Sustain Agric 36:744–758. https://doi.org/10.1080/10440046.2011.627992
Bunch R (1997) Two ears of corn: a guide to people-centered agricultural improvement. World Neighbors, Oklahoma City
Chambers R (1983) Rural development: putting the last first, 1st edn. Pearson Education Limited, London
Chambers R (1992) Rural appraisal: rapid, relaxed and participatory. IDS Discuss Pap 311:1–68
Chambers R (1994) The origins and practice of participatory rural appraisal. World Dev 22:953–969
Chambers R (2007) From PRA to PLA and pluralism: practice and theory. IDS Work Pap 286:1–41
Cleaver F (1999) Paradoxes of participation: questioning participatory approaches to development. J Int Dev 11:597–612. https://doi.org/10.1002/(SICI)1099-1328(199906)11:4<597::AID-JID610>3.0.CO;2-Q
Contreras J (1984) La teoría de la "modernización" y su concepto de cultura campesina: reflexiones críticas. In: Sevilla E (ed) Sobre agricultores y campesinos. Estudios de Sociología Rural en España. Servicio de Publicaciones Agrarias, Madrid, pp 110–148
Cronon W (1996) The trouble with wilderness: or, getting Back to the wrong nature. Environ Hist Durh N C 1:7–28. https://doi.org/10.2307/3985059
Connell JH, Orias E (1964) The Ecological Regulation of Species Diversity. Am Nat 98 (903):399–414. https://doi.org/10.1086/282335
Delgado GC, Cornetta A, Díaz B (2014) Cambio climático global, transformación agraria y soberanía alimentaria en América Latina. CLACSO, CODESRIA, IDEAs, Buenos Aires
Echeverría J (1999) Introducción a la metodología de la ciencia. La filosofía de la ciencia en el siglo 20. Ediciones Cátedra, Madrid
Edwards DP, Gilroy JJ, Woodcock P et al (2014) Land-sharing versus land-sparing logging: reconciling timber extraction with biodiversity conservation. Glob Chang Biol 20:183–191. https://doi.org/10.1111/gcb.12353
Fairhead J, Leach M, Scoones I (2012) Green grabbing: a new appropriation of nature? J Peasant Stud 39:237–261. https://doi.org/10.1080/03066150.2012.671770
Fals-Borda O, Rahman MA (1991) Action and knowledge: breaking the monopoly with participatory action research. Rowman & Littlefield Publishers, Lanham
Feder G, Savastano S (2017) Modern agricultural technology adoption in sub-Saharan Africa: a four-country analysis. In: Pingali P, Feder G (eds) Agriculture and rural development in a globalizing world: challenges and opportunities. Routledge Earthscan, New York, pp 11–25
Feyerabend P (2010) Against method. Verso, London
Foucault M (2008) Seguridad, territorio, población. Akal Ediciones, Madrid
Gibson CC, Ostrom E, Ahn TK (2000) The concept of scale and the human dimensions of global change: a survey. Ecol Econ 32:217–239. https://doi.org/10.1016/S0921-8009(99)00092-0
Gobierno de Canarias (2017) LEY 4/2017, de 13 de julio, del Suelo y de los Espacios Naturales Protegidos de Canarias
Gramsci A (2014) Quaderni dal carcere. Edizione critica dell'Istituto Gramsci, Torino
Grau R, Kuemmerle T, Macchi L (2013) Beyond 'land sparing versus land sharing': environmental heterogeneity, globalization and the balance between agricultural production and nature conservation. Curr Opin Environ Sustain 5:477–483. doi: //dx.doi.org.accedys2.bbtk.ull.es/10.1016/j.cosust.2013.06.001
Han B-C (2016) Die Austreibung des Anderen. S. Fischer Verlag GmbH, Frankfurt
Hardeman E, Jochemsen H (2012) Are there ideological aspects to the modernization of agriculture? J Agric Environ Ethics 25:657–674. https://doi.org/10.1007/s10806-011-9331-5
Henríquez RJ (2013) Psicología de la investigación científica. Escolar y Mayo Editores, Madrid
Huston MA (1994) Biological diversity. The coexistence of species on changing landscape. Cambridge University Press, Cambridge
Huston MA (2014) Disturbance, productivity, and species diversity: empiricism vs. logic in ecological theory. Ecology 95:2382–2396. https://doi.org/10.1890/13-1397.1

Hardin G (1960) The Competitive Exclusion Principle. Science 131 (3409):1292–1297. https://doi.org/10.1126/science.131.3409.1292
Illich I (1969) Celebration of Awareness: A Call for Institutional Revolution. Penguin Books, London
Kapoor I (2001) Towards participatory environmental management? J Environ Manag 63:269–279. https://doi.org/10.1006/JEMA.2001.0478
Kremen C (2015) Reframing the land-sparing/land-sharing debate for biodiversity conservation. Ann N Y Acad Sci 1355:52–76. https://doi.org/10.1111/nyas.12845
Kuhn TS (1970) The structure of scientific revolutions. University of Chicago Press, Chicago
Lakatos I (1970) History of Science and Its Rational Reconstructions. In: Proceedings of the 1970 Biennial meeting of the philosophy of science association. University of Chicago Press, Springer, Boston, pp 91–136
Lewin K (1946) Action research and minority problems. J Soc Issues 2:34–46. https://doi.org/10.1111/j.1540-4560.1946.tb02295.x
MacArthur RH, Wilson EO (1963) An Equilibrium Theory of Insular Zoogeography. Evolution 17 (4):373–387. https://doi.org/10.1111/j.1558-5646.1963.tb03295.x
Martinez-Alier J, Temper L, Del Bene D, Scheidel A (2016) Is there a global environmental justice movement? J Peasant Stud 43:731–755. https://doi.org/10.1080/03066150.2016.1141198
Matijasevic MT, Ruiz A (2013) La construcción social de lo rural. Rev Latinoam Metodol la Investig Soc 5:24–41
Navarro F, Cejudo E, Maroto JC (2014) Reflexiones en torno a la participación en el desarrollo rural: ¿Reparto social o reforzamiento del poder? LEADER y PRODER en el sur de España. EURE 40:203–224. https://doi.org/10.4067/S0250-71612014000300010
Oakley P (1991) Projects with people: the practice of participation in rural development. International Labour Office, Geneve
Pingali PL (2012) Green revolution: impacts, limits, and the path ahead. Proc Natl Acad Sci 109:12302–12308. https://doi.org/10.1073/pnas.0912953109
Ravetz JR (1999) What is post-normal science? Futures 31:647–653. https://doi.org/10.1016/S0016-3287(99)00024-5
Ravetz JR (2006) Post-normal science and the complexity of transitions towards sustainability. Ecol Complex 3:275–284. https://doi.org/10.1016/j.ecocom.2007.02.001
Rogers EM, Svenning L (1969) Modernization among peasants: the impact of communication. Holt, Rinehart and Winston, Inc., New York
Ruiz Rivera N, Galicia L (2016) La escala geográfica como concepto integrador en la comprensión de problemas socio-ambientales. Investig Geogr 89:137–153. https://doi.org/10.14350/rig.47515
Sanders HL, Hessler RR (1969) Ecology of the Deep-Sea Benthos. Science 163 (3874): 1419–1424. https://doi.org/10.1126/science.163.3874.1419
Sánchez de Puerta F (1996) Extensión agraria y desarrollo rural. Sobre la evolución de las teorías y praxis extensionistas. Ministerio de Agricultura, Pesca y Alimentación, Madrid
Sevilla E (2013) Agroecología: Fundamentos del pensamiento social agrario y teoría sociológica. Agroecología 8:27–34
Sevilla Á (2014) Hegemonía, gubernamentalidad, territorio. Apuntes metodológicos para una historia social de la planificación. EMPIRIA Rev Metodol Ciencias Soc 27:49–72. https://doi.org/10.5944/empiria.27.2014.10862
Sousa B (2011) Epistomologías del sur. Utopía y Prax Latinoam Rev Int Filos Iberoam y Teoría Soc 54:17–39
Tax S (1960) Action anthropology. In: Gearing F, Netting RM, Peattie LR (eds) Documentary history of the fox project 1948–1958: a program in action anthropology. University of Chicago Press, Chicago, pp 167–171
Terluin IJ (2003) Differences in economic development in rural regions of advanced countries: an overview and critical analysis of theories. J Rural Stud 19:327–344. https://doi.org/10.1016/S0743-0167(02)00071-2

The European Parliament and of the Council of the European Union (2001) Directive 2001/42/EC of the European Parliament and of the Council of 27 June 2001 on the assessment of the effects of certain plans and programmes on the environment. Official Journal of the European Union

The European Parliament and the Council of the European Union (2000) Directive 2000/60/EC of the European Parliament and of the Council of 23 October 2000 establishing a framework for Community action in the field of water policy. Official Journal of the European Union

The European Parliament and the Council of the European Union (2006) REGULATION (EC) No 1367/2006 OF THE EUROPEAN PARLIAMENT AND OF THE COUNCIL of 6 September 2006 on the application of the provisions of the Aarhus Convention on Access to Information, Public Participation in Decision- making and Access to Justice in Enviro. Official Journal of the European Union

Toledo VM (1990) The ecological rationality of peasant production. In: Altieri MA, Hecht SB (eds) Agroecology and small farm development. CRC Press, Boston, pp 53–60

Valencia M (2009) El concepto desvirtuado de la participación pública medioambiental: una visión jurídica – entretantos. http://www.entretantos.org/el-concepto-desvirtuado-de-la-participacion-publica-medioambiental-una-vision-juridica/. Accessed 11 Feb 2019

Van der Ploeg JD (1990) Sistemas de conocimiento, metáfora y campo de interacción el caso del cultivo de la patata en el altiplano peruano. Agric y Soc:143–166

Van der Ploeg JD (2014) Peasant-driven agricultural growth and food sovereignty. J Peasant Stud 41:999–1030. https://doi.org/10.1080/03066150.2013.876997

Van der Ploeg JD (2015) Peasants and the art of farming. Fernwood Pub, Halifax

West P, Igoe J, Brockington D (2006) Parks and peoples: the social impact of protected areas. Annu Rev Anthropol 35:251–277. https://doi.org/10.1146/annurev.anthro.35.081705.123308

Chapter 10
The Role of Local Perceptions in Environmental Diagnosis

Taline Cristina da Silva, Juliana Loureiro de Almeida Campus, and Regina Célia da Silva Oliveira

10.1 Introduction

The proximity of human populations to natural environments promotes the development of intimate relationships between people and the resources available at these sites (Sieber et al. 2010). Thus, these relationships suggest that human populations have perceptions and a vast traditional ecological knowledge about the historically used resources, which is evidence that human groups may be important allies in studies of nature conservation and environmental diagnoses (Rist et al. 2010; Silva et al. 2016a).

In studies of environmental change, it is important to understand how human groups involved in such changes perceive them, since they are part of ecosystems and also responsible for their changes (Bell 2001). Some authors have accessed local perceptions and verified its participation in changes that involve landscape transformations, increasing or decreasing vegetation cover, changes in land use patterns and the availability of natural resources in general, as well as climate change (Lykke 2000; Xu et al. 2006; Gill and Lantz 2014). Thus, studies that access human perceptions about natural resources are also important to better understand the expectations, satisfactions and dissatisfactions, judgments, and behaviors of human beings toward the environment (Guerra and Abílio 2005).

However, there is some complexity in understanding the aspects of people's perceptions in relation to natural resources, since the very concept of perception can

T. C. da Silva (✉)
State University of Alagoas, Palmeira dos Índios, Brazil

J. L. de Almeida Campus
Embrapa Genetics Resources and Biotechnology, Brasília, Brazil

R. C. da Silva Oliveira
Laboratório de Ecologia e Evolução de Sistemas Socioecológicos (LEA), Federal University of Pernambuco, Recife, Brazil

C. Baldauf (ed.), *Participatory Biodiversity Conservation*,
https://doi.org/10.1007/978-3-030-41686-7_10

have several definitions. For example, Tuan (1980) defines perception as a response of the sense to external stimuli, as a purposeful activity, in which certain phenomena are clearly recorded while others are blocked. According to this author, much of what is perceived has value for the individual, for biological, and for cultural survival. For Silva et al. (2016a), it is difficult to access real local perceptions regarding the environment, since they are influenced by cultural, socioeconomic, and biological factors. Therefore, it is necessary to use appropriate methodological tools (see Silva et al. 2014) and research questions appropriate to the reality of each studied group in order to avoid bias when accessing certain information. Studies have been successful in accessing local perceptions of changes in the abundance of certain resources and their causes (Lykke 2000; Wezel and Lykke 2006), but it is observed, for example, that these have sought information on species of local importance from an economic and/or cultural perspective. Local perceptions on climate change are extremely important, especially when considering the experiences of people who deal directly with the dangers and opportunities arising from a changing climate, but it is also important to consider the complexity that surrounds such phenomena for an adequate methodological design.

Through a compilation of papers and case studies, this chapter seeks to discuss the efficacy of local perceptions in environmental diagnostics, within the context of modified landscapes, climatic changes, and variation in abundance of useful plant resources for these populations, as well as highlighting some aspects of conceptual and methodological approaches that permeate these studies.

10.2 The Importance of Local Perception to Identify Changes at the Landscape Level

According to Fedrizzi and Tomasini (2008), studies related to the environmental perception of human populations have been more intense since the 1960s, a time marked by environmental concerns. Considering that the importance of this type of study has already been emphasized, we will present studies that access local perceptions to understand changes in landscapes while also discussing methodological aspects and risks of bias in these types of approach.

Silva et al. (2017) observed, through biological evidence and local perceptions, that the abundance of useful plant species for the human population around the Araripe National Forest, Northeast Brazil, was modified due to the different forms of landscape management occurring at different moments over time. According to the semi-structured interviews, a historical chart, and floristic survey, the process of landscape management at a given time favored the development of useful heliophilous species, since the local population resided inside the forest and these plant resources underwent incipient domestication. With the creation of the Conservation Unit, families were withdrawn from the interior of the forest, and the process of natural regeneration increased forest cover; however, it has disadvantaged the development of useful heliophilous species, which has consequently diminished their abundance in the landscape.

In Mexico, Dalle et al. (2006) also found changes in the vegetation cover of the Ejido de Quintana Roo Forest, with the help of the local population, through information from participatory mapping. Informants had to indicate the abundance and diversity of the region's plant resources and also indicated their decline over time.

The use of participatory methodologies, an important tool in the studies of rapid environmental diagnoses, is observed in these two studies (see Sieber et al. 2014). However, the complexity of accessing local perceptions has already been mentioned; therefore, we recommend using this methodology concomitantly with others, since at the moment of collectivity there can be several types of influences, such as overlapping voices and gender issues, among others (Evans et al. 2006; Sieber et al. 2014).

Our focus will now be on perception studies regarding landscapes in different socioenvironmental methods and contexts, in order to show the variations in these studies. Tatlıdil et al. (2009) proposed to observe farmers' perceptions through semi-structured interviews, in which they had to indicate, through a Likert scale (a type of rating scale used to measure opinions) from 1 to 5, the degree of importance of the implantation of sustainable agriculture in the landscape of the province of Kahramanmaras (Turkey). However, the use of the Likert scale in some studies should be discussed, since scale values are determined by the researcher, which may bias certain results by overestimating or underestimating some perceptive aspects of the environment.

The work of Gómez-Limón and Fernández (1999) developed in a central region of Madrid (Spain) opted for the use of photographs to verify local perceptions about changes in vegetation. A set of six vegetation photographs in different ecological stages of succession provoked by the agricultural structure in the region were used. Different groups of users (farmers, tourists, and local populations) indicated their preferences for the environments indicated in the photographs. Visual stimuli are important methodological tools used to access local perceptions about resources, since they can stimulate the various senses associated with human perception, but the choice of images is fundamental in this type of study, since images which are distorted, poorly framed, or of familiar places, among other aspects, can generate bias in the results of the study (Ver Silva et al. 2010).

Among the studies on landscape perceptions, it is important to emphasize those conducted with a younger audience, specifically children and adolescents (Barraza and Ceja-Adame 2005; Guerra and Abílio 2005; Bezerra 2006; Schwarz et al. 2007; Silva et al. 2010). For such studies, specific methodological tools adapted for this type of public are needed. Stimulating landscape representation through drawings and questionnaires with appropriate languages for this age group are some of the methodological suggestions.

Data on landscape changes obtained through local perceptions either can aid in accurate environmental diagnoses or may also generate conflicts of interest. For example, Zube (1986) observed that the population living near a National Park in Bali, Indonesia, perceived an increase in conflicts between park managers' interests and the population, which indicated they were excluded from the management plan of these areas, emphasizing once again the importance of local participation from the environmental diagnosis to the management of the areas.

Dhubháin et al. (2009) set out to investigate, through semi-structured interviews, the positive and negative aspects of a reforested area in Ireland through the representation of farmers. The authors restricted themselves to using only one type of tool to access representations. The results were surprising, since most of the interviewees emphasized the negative aspects of reforestation in the area and the authors discussed this result, justifying that there was a complete exclusion of the farmers in the process of implementation of the reforestation project.

10.3 The Importance of Local Perception to Identify Changes in Resource Availability

Several studies have been conducted with the objective of integrating local perception with biodiversity management and management plans (Fraser et al. 2006; López-Hoffman et al. 2006; Gaoue and Ticktin 2009; Fernández-Llamazares et al. 2016; Silva et al. 2016b; Campos et al. 2018). In this chapter, we will discuss the contributions of local populations in studies that seek to understand the changes in the availability of natural resources over time, addressing some case studies, highlighting some methodological biases, as well as the efficiency of such an approach.

The study by Wezel and Lykke (2006) stands out in relation to research related to changes in the availability of plant resources. The authors sought to assess whether local perceptions of changes in abundance of certain plant species in West Africa were similar across three countries of the continent (Burkina Faso, Senegal, and Niger). Seven case studies were evaluated, and the authors found a regional trend of woody species decline, with some differences at the local level, which could be explained by the existence of sacred species and by changes in land use. The changes were measured using data from the perception of farmers and ranchers on the increase or decrease of species managed in vegetation areas, and the survey showed relevant information on species that are threatened with extinction (Wezel and Lykke 2006), generating a list of species.

In Senegal, Lykke (2000) studied the local perception of farmers regarding changes in savanna vegetation within a protected forest. Practically all informants noticed a decline in vegetation density and local extinction of some useful woody species, especially those used for construction. The local population recognized that the scarcity of rains and the increase in fires played a significant role in the decrease of vegetation and mentioned the preference for dense vegetation. Perceptions of local populations were compared with aerial images of the region, which revealed the opening and disappearance of gallery forests, and botanical studies demonstrated that gallery forests were being replaced by African savanna species. The study demonstrated that the need for conservation of some species presented by local populations is aligned with the interests of conservation managers, which may result in the creation of strategies for sustainable use and, consequently, conservation of vegetation.

Kristensen and Balslev (2003) evaluated the local perception of the Gourounsi ethnic group in Burkina Faso regarding the availability of useful woody plants in the local savanna used by the Gourounsi to collect resources related to food, medicine, and construction. The responses varied widely, but most informants said that species availability was high. However, some informants warned that two species used for food are becoming rare, indicating an emerging problem. The authors also sought to investigate whether the frequency of visits to the savanna was correlated with their perception of the availability of species, hoping that those who visit the savanna with less frequency would indicate a high availability, which was not the case. The authors concluded that local vegetation was not being degraded rapidly and that forest products were supplied in sufficient quantities; however, the scenario was changing and a warning about the need to conserve local resources was needed.

In the Bolivian Amazon, Fernández-Llamazares et al. (2016) sought to assess whether there was an alignment between the local perception of the Tsimane indigenous group and ecological data regarding the availability of the palm tree *Geonoma deversa* (Poit.) Kunth. Moreover, the researchers looked at whether the different perceptions regarding changes in the abundance of species populations explained the collection behavior and the local management practices. The results showed that local perceptions were partially aligned with ecological estimates and that the local perception from villages near the commercial center, where the populations of *G. deversa* are distant from residences, aligned with the ecological data with relative frequency. On the other hand, the Tsimane who lived in villages distant from the commercial center underestimated the availability of *G. deversa*. Fernández-Llamazares et al. (2016) also found that the collection and management practices performed by the Tsimane in relation to the palm species were shaped by the local representation regarding the availability of the species. Human perceptions varied according to the cultural, economic, and social conditions that the Tsimane had experienced, evidencing the complex nature of local perceptions (Fernández-Llamazares et al. 2016).

Along the same line of research, Gaoue and Ticktin (2009) observed, through an ethnoecological study with the Fulani in Benin, Africa, that the perception of this group was highly convergent with the scientific data related to *Khaya senegalensis* (Desv.) A. Juss., a species whose leaves were collected by the Fulani for use as fodder. However, the Fulani perceptions regarding the main threats to the populations of *K. senegalensis* differed significantly among the Fulani from two different regions, which mentioned different reasons for the decrease in the population of the species. The authors discussed the question of the sustainable management of commonly used resources, located on public lands and that are used by different social groups. In the case of *K. senegalensis*, the extraction is carried out by farmers, loggers, and the Fulani. Each group from this socioecological system tends to maximize its use, which may cause a local extinction (Gaoue and Ticktin 2009). For the authors, under commonly used conditions, accessing local perceptions of threats to these resources may be a better predictor than local ecological knowledge related to species use patterns.

Also studying a commonly used resource, Campos et al. (2018) sought to verify local perceptions of the Fulni-ô indigenous people in Pernambuco, Northeast Brazil, regarding changes in the abundance of the populations of *Syagrus coronata* (Mart.) Becc, a species whose leaves are extracted for the production of handicrafts by the Fulni-ô. Moreover, factors that influenced the sustainability of the collection practice were investigated. The authors found that the most experienced extractors performed the collection in a more sustainable manner compared to the less experienced. Although the Fulni-ô pointed out that the main reason for the resource's decrease was the change in land use generated by the leasing of land, they did not recognize themselves as causing the decrease in the resource. The authors argue that the perception of resource scarcity alone does not mean that conservation strategies will be implemented easily and that it is important to identify which factors are perceived as threats by groups that use the resource. In addition, the authors acknowledge that the lack of ecological data in the past may have limited the discussion of resource sustainability, which has been strengthened by the use of participatory methods that analyzed local perceptions about populations of *S. coronata* in past decades, such as a historical chart (Campos et al. 2018).

In Venezuela, López-Hoffman et al. (2006) found that the extractivists' perception of sustainable collection rates of the species *Rhizophora mangle* L. did not coincide with the levels of sustainable collection according to the population ecology of the species. The authors observed that the older extractivists perceived a larger population decline in *R. mangle* and carried out a less intensive extraction when compared to younger extractivists. In the region, there are no other species that can replace its collection for use in the timber industry. López-Hoffman et al. (2006) believe that the absence of alternative logging resources and the different socioeconomic conditions of extractivists may have reflected their perception of sustainability, creating a trade-off between the need for mangrove preservation and the economic necessity of collection. The authors recommend that the collection regime be limited and adapted to the ecological response of *R. mangle* populations. Thus, they suggest that extractivists participate as monitors in community-based management programs.

Continuing with studies on the local perception of a particular plant species, Ekue et al. (2010) found the perception of the local population in East Africa regarding the importance of the tree *Blighia sapida* K. Konig. They observed that the studied group perceived different phenotype varieties of the species, and this species is in the initial process of being domesticated.

Consequently, there is a debate within the scientific literature regarding the effectiveness of access to the perceptions of local populations as a guide in the management of natural resources. Many authors recommend that perceptions about the availability, uses, and management practices of species extracted by human populations be included in research that seeks to verify the sustainability of extractive actions, since this type of study can contribute to the creation of management plans for the species together with the collectors (Xu et al. 2006; Rist et al. 2010).

However, in some cases, it is possible to demonstrate that the perception of human populations on the availability of resources often does not coincide with what is measured by researchers (see López-Hoffman et al. 2006), indicating the need to understand the scenario that surrounds this relationship so that the implementation of conservation strategies is achieved with success.

The divergence of information between local perceptions and ecological surveys about the availability of a given resource may occur due to methodological biases of the surveys. For example, the perception of resource availability is strongly associated with the uses related to it, that is, if the used part of a given plant is the leaf, when asked about its availability, the answers will possibly be associated with the amount of leaves available, while the ecological data may consider the number of individuals in the species population. Therefore, information will not always be compatible with reality, depending on the accessed data. Additionally, it is important to highlight that perceptions are influenced by cultural, psychological, and socioeconomic factors (Silva et al. 2016b). The work of Silva et al. (2011) observed divergences in the perceptions regarding the availability of a set of useful tree species in a riparian vegetation. This divergence was possibly due to the participation of different social actors in the study (farmers, agricultural workers, traditional communities) that, consequently, had different degrees of schooling, income, and age and that also made different uses of the vegetation.

10.4 The Importance of Local Perception Studies to Identify Climate Change

Global climate change is considered one of the most important political, scientific, environmental, and social challenges at the global level due to the frequency and intensity that its consequences have been affecting not only the climate but also the natural environment and human development (IPCC 2013; 2014; Sisifa et al. 2016). Thus, in order to better understand the extent of these impacts, the scientific literature has also emphasized the importance of knowing how human populations have been perceiving, interpreting, and dealing with the impacts of climate change at a local scale (Campos et al. 2014; Fernandez-Llamazares et al. 2015; Silva et al. 2016a; Oliveira et al. 2017). Thus, in this topic we will address aspects related to research on climate perception and local adaptive actions, and we will also highlight the main biases and suggestions. Moreover, there are some terms that are specific to this topic and can be consulted in Table 10.1.

Using this perspective, Fernandez-Llamazares et al. (2015) described the ethnoclimatic knowledge of human populations and investigated the effect of external information on the perception of climate change of indigenous peoples living in the Bolivian Amazon. This research found that not all Tsinames attributed negative significance to climate disturbances, since for many changes to the climate brought local benefits. Furthermore, the study also demonstrated that even when providing scientific

Table 10.1 Some concepts applied by studies that involve perception related to climate change

Term	Concept	Reference
Global environmental changes	Changes in the physical and biochemical environment, that is, they include losses/gain in biodiversity, changes in soil and water quality, and so on	Leemans et al. (2009, Wolverton et al. (2014)
Global climate changes	This event is related to significant changes in the average meteorological conditions or a variation in climate that has occurred over a long period of the history of earth. Changes in seasons include, but are not limited to, increase in temperature and extreme events, such as droughts and floods	IPCC (2014), WMO (2014)
Risk perception	Can be understood as synonymous with negative judgment, danger, or problem that people attribute to a perceived environmental event, that is, adverse situation to the human way of life and ecosystems	Smith et al. (2000), Granderson (2014), Oliveira et al. (2017)
Adaptive strategy/adaptive behavior	Refers to the human capacity to adjust to environmental changes, including opportunities and/or minimization of risks	Smit and Wandel (2006), IPCC (2014), Oliveira et al. (2017)
Exhibition	Can be defined as the magnitude that an environmental impact influences the environment and people's way of life in a negative way	IPCC (2014)
Sensitivity	Is understood as the degree to which the socioecological system is affected by environmental changes, from either its negative effects or those that are seen as local benefits	IPCC (2014)
Vulnerability	Includes exposure and human sensitivity and ecosystems to a potential impact. In the context of climate change, it characterizes the lack of effective action to address and/or adapt to the confronted risks	Adger et al. (2009), IPCC (2014), Sisifa et al. (2016)

data on the impacts of climate change, some informants were skeptical of the climate data presented by the researchers, as they did not agree with what was observed locally. This means that there may be divergences between knowledge of climate change and scientific data. Therefore, for a better dialogue between scientific and local communities on impacts and adaptations to a changing climate, it is first necessary to understand how these events are being processed, interpreted, and translated by different social actors in the same community (Fernandez-Llamazares et al. 2015).

In Tucuiz, Mexico, researchers investigated how different groups of farmers perceived and adapted to an increase in climatic variability (Campos et al. 2014). Campos et al. (2014) observed that small farmers had various strategies for adapting to climate impacts and these actions were being based on the perception of value and usefulness of local landscapes. In Tucuiz the younger and more educated residents were also the ones who developed the best means to direct their cultivations even in the face of climatic disturbances. In contrast, in the same community, there were social groups that only waited for the rain to continue subsistence agriculture, because they experienced greater poverty; they were also the most vulnerable to climatic impacts that affected the community. However, the study found that those who were able to adapt had perceptions based on climate risk, and this observation provided resilience to some groups as they began to incorporate efficient actions in agriculture to deal with local climate change (Campos et al. 2014).

It is also possible to observe the influence of other socioeconomic factors (besides income, age, and schooling seen in the previous study) in Jones and Boyd (2011). Although the studies are from different research groups and different regions of Africa, the authors found a similar reality among the study participants. The female participants were identified as more exposed and vulnerable to climatic hazards, even when they had more knowledge and greater environmental concerns than the male participants. Jost et al. (2016) believe that this fact can be better explained if sociocultural rules are considered. Thus, the authors suggest that when finding a similar reality, it is important to recognize that such phenomenon may be more related to the role that the individual plays in society, rather than gender. Therefore, in general, women have their responsibilities linked to domestic activities, and devoting themselves to agricultural work and/or migrating to other regions in search of employment may be offered advantages to men in the face of climatic disturbances (Jost et al. 2016).

In a literature review, Granderson (2014) noted that only examining individual perceptions and actions attributed to environmental risks offers a good scientific response; nevertheless, it is an incomplete response. Therefore, it is important to understand what knowledge, value (material and symbolic), and strategies are specific to each local reality. This will provide better insights into the actual process of building adaptive capacity to react and/or adapt to climate change (Granderson 2014). The author further suggests that adopting an approach that is designed to analyze cultural and political dimensions will provide a more robust response about whether local decisions are tied to specific interests, as well as understanding "which" and "why" these factors are dominating decisions of local risk.

When considering cultural aspects related to risk perception, Oliveira et al. (2017) analyzed how religious and spiritual aspects directed observations and adaptive strategies in the context of climatic uncertainties faced by residents of the rural community of Pernambuco, Northeast Brazil. The authors observed that dimensions of religiosity/spirituality, such as religious history, values/beliefs, commitment, and daily spiritual experiences, exert both negative and positive influence on the richness and sharing of perceived risks, as well as locally known adaptive strategies (Oliveira et al. 2017). Although this approach offers advantages, researchers still suggest that it is important to note that more robust results that allow for generalizations (and avoid biases that may be related to specific religious groups) require a more religiously diverse setting.

Given the above, studies that aim to better understand global climate change have increased, and such knowledge has been shown to be more effective when evaluated at the community level, since for both theoretical and practical understandings, public policies will be more effective when local, regional, and global information is available (IPCC 2013, 2014; Granderson 2014). Moreover, many authors argue that there are still many gaps in knowledge, given the complexity involved in both human perceptions and behaviors, as well as climate phenomena itself (Granderson 2014; Fernandez-Llamazares et al. 2015). Intrinsic aspects of the human beings associated with the fact that each person lives under different social, economic, political, cultural, and environmental conditions will further affect views on the world, leading to the distortion of meanings and responses to climatic problems faced in the same community (Gill and Lantz 2014).

Thus, understanding how local people perceive global climate change and its impacts is useful to generate new scientific, social, and political discussions aimed at effective mitigation of the disruption caused by the phenomenon and the risks from the environmental event.

10.5 Final Considerations

With the present chapter, we sought to show the reader the importance of accessing the perception of local populations to answer questions that involve landscape modifications and changes in resource availability. We also emphasize the role of local perceptions in the identification of adaptations and strategies of human groups in the face of events that involve climate change. Understanding the particularities of the studied group and using appropriate methodologies to investigate how these groups perceive the consequences of their own actions on the used resources and managed landscapes are of paramount importance for the implementation of conservation measures to be successfully achieved. Furthermore, accessing the factors behind how people perceive the environment and climate change can help in understanding the cultural and biological filters that influence different perceptions. Understanding these issues can help in comprehending human behavior related to how people appropriate natural resources, contributing to the conservation of these resources, as well as promoting risk minimization in a scenario of global environmental change.

References

Adger WN, Dessai S, Goulden M et al (2009) Are there social limits to adaptation to climate change? Clim Change 93:335–354

Barraza L, Ceja-Adame MP (2005) Los niños de la comunidad: su conocimiento ambiental y su percepcion sobre "Naturaleza". México. disponível em: http://www.ine.gob.mx/ueajei/publicaciones/libros/420/dieciseis.html. Accessed 17 June 2007

Bell S (2001) Landscape pattern, perception and visualisation in the visual management of forests. Landscape Urban Plan 54:201–211

Bezerra TMO (2006) Percepção do ambiente por alunos e professores no entorno da Estação Ecológica de Caetés, Paulista, Pernambuco. 50 f. Dissertation, Universidade Federal Rural de Pernambuco, Recife

Campos M, Velázquez A, Mccall M (2014) Adaptation strategies to climatic variability: a case study of small-scale farmers in rural Mexico. Land Use Policy 38:533–540

Campos JLA, Araújo EL, Gaoue OG et al (2018) How can local representations of changes of the availability in natural resources assist in targeting conservation? Sci Total Environ 628:642–649
Dalle SP, Blois S, Caballero J et al (2006) Integrating analyses of local land-use regulations, cultural perceptions and land-use/land cover data for assessing the success of community-based conservation. For Ecol Manag 222:370–383
Dhubháin ÁN, Fléchard M, Moloney R et al (2009) Stakeholders' perceptions of forestry in rural areas-two case studies in Ireland. Land Use Policy 26:695–703
Ekue MR, Sinsi B, Eyogi-Matig O et al (2010) Use, traditional management, perception of variation and preferences in Ackee (*Blighia sapida* K.D Koenig) fruit traits in Benin: implication for domestication and conservation. J Ethnobiol Ethnomed 6:12
Evans K, Jong WD, Cronkleton P et al (2006) Guide to participatory tools for forest communities. Bogor, CIFOR, p 37
Fedrizzi B, Tomasini SLV (2008) Projetando ambientes mais sustentáveis com a elaboração da psicologia ambiental. In: Pinheiro JQ, Gunther H (Org). Método de pesquisa nos estudos Pessoa-Ambiente, São Paulo: Casa do psicólogo 1: 313–342
Fernandez-Llamazares A, Luz A, Cabeza C et al (2015) Rapid ecosystem change challenges the adaptive capacity of local environmental knowledge. Global Environ Chang 31:272–284
Fernández-Llamazares Á, Díaz-Reviriego I, Guèze M et al (2016) Local perceptions as a guide for the sustainable management of natural resources: empirical evidence from a small-scale society in Bolivian Amazonia. Ecol Soc 21(1)
Fraser DJ, Coon T, Prince MR et al (2006) Integrating traditional and evolutionary knowledge in biodiversity conservation: a population level case study. Ecol Soc 11(2):4
Gaoue OG, Ticktin T (2009) Fulani knowledge of the ecological impacts of *Khaya senegalensis* (Meliaceae) foliage harvest in Benin and its implications for sustainable harvest. Econ Bot 63:256–270
Gill H, Lantz T (2014) A community-based approach to mapping Gwich'in observations of environmental changes in the lower Peel River watershed, NT. J Ethnobiol 34(3):294–314
Gómez-Limón J, Fernández JVL (1999) Changes in use and landscape preferences on the agricultural-livestock landscapes of the Central Iberian Peninsula (Madrid, Spain). Landscape and Urban Plan 44:165–175
Granderson AA (2014) Making sense of climate change risks and responses at the community level: a cultural-political lens. Clim Risk Manag 3:55–64
Guerra RAT, Abílio FJP (2005) A percepção ambiental de professores de escolas públicas de ensino fundamental de Cabedelo, Paraíba. In: Abílio FJP, Guerra RAT (eds) A Questão Ambiental no Ensino de Ciências: A formação continuada de professores de ensino fundamental, vol 1. UFPB/FUNAPE/LEAL, João Pessoa, pp 91–104
IPCC – Intergovernmental Panel on Climate Change (2013) Climate change: the physical science basis. Contribution of working group I to the fifth assessment report of the intergovernmental panel on climate change
IPCC – Intergovernmental Panel on Climate Change (2014) In: Pachauri RK, Meyer LA (eds) Climate change 2014: synthesis report. contribution of working groups I, II and III to the fifth assessment report of the intergovernmental panel on climate change, Genebra, p 151
Jones L, Boyd E (2011) Exploring social barriers to adaptation: insights from Western Nepal. Global Environ Chang 21:1262–1274
Jost C, Kyazze F, Naab S et al (2016) Understanding gender dimensions of agriculture and climate changein smallholder farming communities. Climate and development 8(2):133–144
Kristensen M, Balslev H (2003) Perceptions, use and availability of woody plants among the Gourounsi in Burkina Faso. Biodivers Conserv 12(8):1715–1739
Leemans R et al (2009) Developing a common strategy for integrative global environmental change research and outreach: the Earth System Science Partnership (ESSP). Curr Opin Environ Sustain 1:4–13

López-Hoffman L, Monroe IE, Narváez E et al (2006) Sustainability of mangrove harvesting: how do harvesters' perceptions differ from ecological analysis? Ecol Soc 11(12):14

Lykke AM (2000) Local perceptions of vegetation change and priorities for conservation of woody-savanna vegetation in Senegal. J Environ Manag 59(2):107–120

Oliveira RCS, Albuquerque UP, Silva TLL et al (2017) Religiousness/spirituality do not necessarily matter: effect on risk perception and adaptive strategies in the semi-arid region of NE Brazil. Global Ecol Conserv 11:125–133

Rist L, Shaanker RU, Milner-Gulland EJ et al (2010) The use of traditional ecological knowledge in forest management: an example from India. Ecol Soc 15(3)

Schwarz ML, Sevegnani L, André P (2007) Representações da Mata Atlântica e de sua biodiversidade através dos desenhos infantis. Rev Brasileira de Biociências 5:744–746

Sieber SS, Medeiros PM, Albuquerque UP (2010) Local perception of environmental change in a semi-arid area of Northeast Brazil: a new approach for the use of participatory methods at the level of family units. J Agr Environ Ethics 24(5):511–531

Sieber SS, Silva TC, Campos LZO et al (2014) Participatory methods in ethnobiological and ethnoecological research. In: Albuquerque UP, LVFC C, Lucena RFP et al (eds) Methods and techniques in ethnobiology and ethnoecology. Springer, Recife, pp 39–48

Silva TC, Medeiros PM, Araújo TAS et al (2010) Northeastern Brazilian students' representations of Atlantic Forest fragments. Environ Develop Sustain 12:195–211

Silva TC, Ramos MA, Alvarez IA et al (2011) Representações dos proprietários e funcionários de fazendas sobre as mudanças e conservação da vegetação ciliar às margens do rio São Francisco, Nordeste do Brasil. Sitientibus 11:279–285

Silva TC, Cruz MP, Araújo TAS et al (2014) Methods in research of environmental perception. In: Albuquerque UP, Cunha LVFC, Lucena RFP et al (eds) Methods and techniques in ethnobiology and ethnoecology. Springer, Recife, pp 39–48

Silva TC, Chaves LS, Albuquerque UP (2016a) What is environmental perception? In: Albuquerque UP, Alves RRN (eds) Introduction of ethnobiology. Springer, Heidelberg, pp 93–98

Silva TC, Medeiros MFT, Peroni N (2016b) Folk classification as evidence of transformed landscapes and adaptative strategies: a case study in the semiarid region of northeastern Brazil. Landsc Res 9:1–13

Silva TC, Campo LZO, Balée W et al (2017) Human impact on the abundance of useful species in a protected area of the Brazilian Cerrado by people perception and biological data. Landsc Res 13:1–14

Sisifa A, Taylor M, McGregor A et al (2016) Pacific communities, agriculture and climate change. In: Taylor M, Mcgregor A, Brian D (eds) Vulnerability of Pacific Island agriculture and forestry to climate change. Pacific Community (SPC), p 551

Smit B, Wandel J (2006) Adaptation, adaptive capacity and vulnerability. Glob Environ Chang 16(3):282–292

Smith K, Barrett CB, Box PW (2000) Participatory risk mapping for targeting research and assistance: with an example from East African pastoralists. World Dev 28(11):1945–1959

Tatlıdil FF, Boz I, Tatlıdil H (2009) Farmers' perception of sustainable agriculture and its determinants: a case study in Kahramanmaras province of Turkey. Environ Develop Sustain 11:1091–1106

Tuan Y (1980) Topofilia um estudo da percepção, atitudes e valores do meio ambiente. Difel, Difusão editorial S.A., São Paulo, p 288

Wezel A, Lykke AM (2006) Woody vegetation change in Sahelian West Africa: evidence from local knowledge. Environ Develop Sustain 8:553–567

WMO – World Meteorological Organization (2014) Weather report for 2050 in Brazil highlights impact of climate change. https://public.wmo.int/en/media/news/weather-report-2050-brazil-highlights-impact-of-climate-change. Accessed 14 June 2018

Wolverton SK, Chambers JE, Veteto JR (2014) Climate change and ethnobiology. J Ethnobiol 34:273–275

Xu J, Chen L, Lu Y et al (2006) Local people's perceptions as decisions support for protected area management in Wolong Biosphere Reserve, China. J Environ Manag 78:362–372

Zube EH (1986) Local and extra-local perceptions of national parks and protected areas. Landsc Urban Plan 13:11–17

Chapter 11
Participation in Biocultural Diversity Conservation: Insights from Five Amazonian Examples

Álvaro Fernández-Llamazares, Petra Benyei, André B. Junqueira, and Victoria Reyes-García

11.1 Introduction

Just as the biosphere is being severely eroded by global change, so is the ethnosphere and probably at greater rates (Ferguson and Messier 1997; Cox 2000; Brodt 1999; Godoy et al. 2005; Brosi et al. 2007; Turner and Turner 2008; Reyes-García et al. 2007, 2013; Tang and Gavin 2016; Gavin et al. 2015, 2018). Indeed, some researchers argue that the losses of biological and cultural diversity are inextricably linked and driven by the same threats and pressures (Maffi 2005; Pretty et al. 2009; Gorenflo et al. 2012; Rozzi 2012). In response to this, a growing body of research and policy initiatives have adopted biocultural approaches to conservation (sensu Gavin et al. 2015). These approaches rest on the idea that the conservation of a substantial proportion of the world's biodiversity largely depends on Indigenous and Local Knowledge (hereinafter ILK), or the knowledge, practices, and beliefs of Indigenous Peoples and Local Communities (Brondizio and Le Tourneau 2016; Garnett et al. 2018). In parallel to the wide array of methods developed to conserve and manage biodiversity from the bottom-up, the past three decades have witnessed

Á. Fernández-Llamazares (✉)
Helsinki Institute of Sustainability Science (HELSUS), Faculty of Biological and Environmental Sciences, University of Helsinki, Helsinki, Finland

Global Change and Conservation (GCC), Organismal and Evolutionary Biology Research Programme, University of Helsinki, Helsinki, Finland
e-mail: alvaro.fernandez-llamazares@helsinki.fi

P. Benyei · A. B. Junqueira
Institut de Ciència i Tecnologia Ambientals (ICTA), Universitat Autònoma de Barcelona, Bellatera, Barcelona, Spain

V. Reyes-García
Institut de Ciència i Tecnologia Ambientals (ICTA), Universitat Autònoma de Barcelona, Bellatera, Barcelona, Spain

Institució Catalana de Recerca i Estudis Avançats (ICREA), Barcelona, Spain

C. Baldauf (ed.), *Participatory Biodiversity Conservation*,
https://doi.org/10.1007/978-3-030-41686-7_11

the emergence of myriads of initiatives focused on ILK conservation, revitalization, protection, documentation, and/or maintenance all over the world (Aikenhead 2001; Gavin et al. 2015; Fernández-Llamazares and Cabeza 2017).

Three recent systematic reviews have identified five non-exclusive types of initiatives oriented to ILK conservation (McCarter et al. 2014; Tang and Gavin 2016; Benyei et al. 2019): (1) *community-based ILK conservation initiatives* such as those protecting and promoting traditional lifeways and/or the commercialization of ILK-based products at the community level (Little 2005; Klein 2011; Simpson et al. 2013); (2) *capacity-building initiatives* aiming at strengthening IPLCs' alliances and financial autonomy to confront ILK misappropriation, contributing to its protection and maintenance (Maikhuri et al. 2005; Subba Rao 2006); (3) *education and awareness efforts* such as customary education programs that integrate ILK in school curricula, contributing to strengthen ILK transmission (Kimmerer 2002; Castagno and Brayboy 2008; Ruiz-Mallén et al. 2010; McCarter and Gavin 2011, 2014; Hamlin 2013; Abah et al. 2015); (4) *policy and legislation initiatives* in which the need to preserve and integrate ILK in conservation is acknowledged and/or enforced through policy or by law (Alexander et al. 2004); and (5) *research and documentation projects* focusing on the compilation of ILK in databases and registers for its protection (Gadgil et al. 2000; Pardo-de-Santayana 2014; Bussmann et al. 2018).

Although these initiatives are as diverse as the locations and IPLCs they emanate from, they generally offer numerous opportunities for strengthening customary institutions for ecosystem management, biodiversity conservation, and ecological restoration (McCarter and Gavin 2014; López-Maldonado and Berkes 2017; Ban et al. 2018; Fernández-Llamazares and Cabeza 2017; Reyes-García et al. 2018). However, participation of IPLCs in ILK conservation is a major task ahead. For example, a recent study shows that IPLCs are rarely participating in more than one phase of the ILK conservation initiatives (Benyei et al. 2019). The reasons for this lack of participation are still not clear, and thus, there have been calls (1) to enrich the analysis of the degree of IPLC participation in conservation initiatives (Tang and Gavin 2016) and (2) expand the types of evidence assessed in reviews to more fully and rigorously integrate rich and multifaceted qualitative insights (Sterling et al. 2017a).

With these goals in mind, in this chapter, we examine the participation of IPLCs in ILK conservation through an in-depth analysis of five initiatives in the Amazon Basin. Each of these initiatives was selected to characterize each of the five types of ILK conservation described above. Initiatives were chosen on the basis of availability of academic and gray literature describing them in detail. Complementing previous studies taking a more quantitative approach (e.g., Aswani et al. 2018; Benyei et al. 2019), here we prioritize a qualitative appraisal framework assessing the participation of IPLC in each ILK conservation initiative. While the in-depth analysis of these cases might provide some insights on factors enabling and challenging IPLC participation in ILK conservation initiatives, we are aware they do not represent the entire spectrum of all the existing initiatives in this vein. Although for one of the case studies (i.e., the Tsimane' case) we speak from a more insider perspective (as we have been involved in some of the phases related to this initiative), our analysis is mostly based on literature review. We thus stress that we do not speak on behalf of any of the projects described in this chapter, but rather assess them based on the literature.

While ILK conservation initiatives are found in many areas of the world (Tang and Gavin 2016), we focus on initiatives on the Amazon Basin for two main reasons. First, with over 300 Indigenous groups and more species of plants and animals than any other terrestrial ecosystem in the planet, Amazonia is largely considered as a global hotspot of both biological and cultural diversity (Hoorn et al. 2010; Le Tourneau 2015). Second, IPLC rights and livelihoods have been under threat since European arrival, but in spite of the initiatives mentioned above, these are escalating all over the Amazon owing to the recent sociopolitical instability in the region (Escobar 2018; Artaxo 2019; Codato et al. 2019). For instance, pledges by the Bolsonaro Government in Brazil to open Indigenous lands to mining, agri-business, and infrastructure development represent a direct threat to many IPLCs and their knowledge systems (Begotti 2019). Given the crucial role of IPLCs in conserving and managing some of the most biodiverse landscapes in the region (e.g., Nolte et al. 2013; Blackman et al. 2017; Schleicher et al. 2017), there is a greater need as ever to critically evaluate IPLC participation in ILK conservation initiatives in the Amazon, thereby improving our understanding of the conceptual, procedural, and normative underpinnings of biocultural conservation efforts.

11.2 Case 1: Community-Based ILK Conservation – Basket-Weaving Programs Among the Kaiabi

The first initiative selected focuses on a long-term community-based project developed for the revitalization of weaving knowledge among three Kaiabi (also known as Kawaiwete) Indigenous groups in the Brazilian Amazon. The Kaiabi are a Tupi-Guarani-speaking people who originally occupied several tributaries of the Tapajós River in the southern Brazilian Amazon. Between the 1950s and the 1960s, the Brazilian federal government led the relocation of most of the group to the southeast, in an area that is currently known as the Xingu Park (Grünberg 2004). The resettled Kaiabi have undergone a process of social, institutional, and political innovation and have maintained a repertoire of traditional practices, knowledge, and institutions (Athayde et al. 2009; Athayde and Schmink 2014). Weaving is considered as an important cultural practice by the Kaiabi, and the graphic designs represented in basketry and textiles are strong symbols of their cultural identity (Athayde et al. 2009), reflecting aspects of the group's history, cosmology, ecology, and socioeconomic organization (Athayde et al. 2017a).

There is well-established evidence that several important elements of basket-weaving knowledge among the Kaiabi have eroded over the past decades (Athayde et al. 2017b). The relative unavailability of one of the main natural fibers used in the weaving work (*Ischnosiphon gracilis*) is considered as an important driver of knowledge loss (Athayde et al. 2006). This problem was initially identified by Kaiabi leaders, concerned about the erosion of basket-weaving knowledge among men and women (Athayde et al. 2017a). Aiming to revitalize weaving knowledge, this group of Kaiabi leaders developed a "community-based project for cultural revitalization" (see Athayde et al. 2017b, pp. 535) named Kaiabi Araa ("Design of the Kaiabi"). The project was planned and executed by four Kaiabi Indigenous Communities (Athayde et al. 2017b), as part of the Xingu Program of the Instituto Socioambiental (ISA) and in partnership with the Indigenous organizations Associação Terra Indígena Xingu (ATIX) and Kaiabi Association in the Teles Pires (Kawaip). It was funded by the Indigenous Peoples Demonstrative Projects (PDPI) from the Pilot Program to Conserve the Brazilian Rain Forest (PPG7).

The project lasted 7 years and included different activities such as weaving workshops, field trips, and ecological management of the plants whose fibers are used in basketry (Athayde et al. 2006, 2009). A total of 67 people participated in the several Indigenous-led workshops. Through this project, several weaving workshops were organized to revitalize weaving knowledge of baskets and graphic designs among the group (Athayde et al. 2017a). These workshops prioritized many-to-many knowledge transmission as a learning model, where many elders or teachers taught many apprentices in a spirit of collaborative learning (Athayde et al. 2017b), including transmission of weaving techniques across genders, explicitly recognizing the crucial role of women in safeguarding weaving knowledge (Athayde and Silva-Lugo 2018). The project also included transplanting experiments to regrow the main natural fibers used in Kaiabi basketry (Athayde and Silva-Lugo 2018), as well as the search for substitute natural fibers to use in basketry (Athayde et al. 2006). In addition to these ILK revitalization workshops, other outcomes of the project included the production of educational materials and a participatory video documentary on Kaiabi basket-weaving knowledge (see Athayde et al. 2017b). The video-documentary was one of the winners of the "Indigenous Cultures" award from the Ministry of Culture in Brazil in 2007, further helping to recognize the cultural value of Kaiabi basket-weaving knowledge at the national level (Athayde et al. 2017b).

One of the problems identified by Kaiabi leaders from the onset was that some basket-weaving designs were being rapidly forgotten (Athayde et al. 2017a). In partnership with local researchers, the Kaiabi leaders decided to contact several museums, libraries, and ethnographic collections to request the repatriation of several Kaiabi graphic designs that had been documented in the 1960s (Athayde et al. 2017a). These designs, which included the "ta'agap" (mythical figure), were collected and returned to the communities in both printed and digital formats and have been compiled as part of a book on Kaiabi basketry (Athayde 2006). This book is being used to teach some of the designs that were being lost (Athayde et al. 2017b). To assess the direct impact of the Kaiabi Araa project on the basket-weaving knowledge, a comparative longitudinal assessment of knowledge dynamics was conducted before and 5 years after the Kaiabi Araa cultural revitalization project was developed. The results of this study show, among other things, that the project had a significant effect on the number of basketry designs known by project participants (Athayde et al. 2017a).

11.3 Case 2: Indigenous Capacity-Building – The COICA Alliance

Our second case study refers to the Coordinating Body for the Indigenous Peoples' Organizations of the Amazon Basin (COICA), a capacity-building initiative focused on promoting alliances among IPLCs in the Amazon Basin so as to strengthen their collective capacity to advocate for their rights, including the right over their ILK (Jacanamijoy Tisoy 2011). The COICA coordinates nine Indigenous organizations that represent around 400 Indigenous communities and an estimated population of 1.5 million people (Jacanamijoy Tisoy 2011). Funded in Lima in 1984, it now coordinates the Indigenous Peoples' organizations of all the countries that make up the Amazon Basin. These organizations represent communities from a wide range of settings and with diverse degrees of integration into the market economy and ILK erosion (Loh and Harmon 2005; Gorenflo et al. 2012).

Since its foundation, the COICA has been present as an advocacy and negotiating stakeholder in many of the international discussions on biodiversity conservation and Indigenous Peoples' rights. The COICA had an important role in the negotiation of Convention 169 of the International Labour Organization (ILO) on Indigenous Peoples' right to self-determination and the Article 8(j) of the CBD on the recognition and protection of ILK (Jacanamijoy Tisoy 2011; Varese 1995; Mato 2000). Regarding alliance and partnership building as a strategy for ILK conservation, one of the most relevant COICA-led actions has been the organization of the Amazonian Summits (*Cumbres Amazónicas*). These international meetings of Indigenous Peoples and environmentalist organizations were initiated in 1990 and replicated in 2011, 2013, 2016, and 2018, bringing together all the member organizations, as well as partners, to debate on issues that range from land rights to climate change or pollution generated by extractive industries (Mato 2000; COICA 2011, 2013, 2016, 2018). Although ILK conservation has not been the main focus of any of these meetings, the meetings have proven to be effective for the strengthening of Indigenous networks and the production of declarations that are a powerful way of collectively denouncing violations of Indigenous rights and propose consensual actions to confront them (Herrera 2016). Moreover, some declarations have specific sections on the importance of protecting and maintaining ILK as basic for the Indigenous lifeways and economies under the *Vida Plena* paradigm. More specifically the Manaus declaration (COICA 2011) proposed securing ILK and preventing its unrightful appropriation and commercialization as a key action to be promoted.

The COICA board is formed by representatives of all the organizations and countries, for which the initiative seems to be inclusive of the different regional perspectives overcoming past communication and hierarchical issues reported

(Varese 1995). However, there has been a recent call from youth sectors of the Indigenous organizations to be further included in the COICA board (Comunicaciones COICA 2018). Finally, COICA's foundation and functioning is somewhat reliant on external funding from Western NGO's, specifically Intermon Oxfam which founded the inception meeting in 1984 (Mato 2000; Herrera 2016). This means that despite the high level of Indigenous control over this initiative, its financing is not fully in the hands of IPLCs.

11.4 Case 3: Education and Awareness – Tsimane' Educational Programs

The coordinated work of a partnership of researchers and Tsimane' Indigenous peoples over almost two decades constitutes an example of an education and awareness building ILK conservation initiative. The Tsimane' are a population of hunter-horticulturalists who live in a territory mostly covered by *terra firme* lowland rainforests, extending from the Andean piedmont to the savannas of Moxos in the Department of Beni, in the Bolivian Amazon (Paneque-Gálvez et al. 2013; Guèze et al. 2015). The Tsimane' number approximately 14,000 people living in about 125 villages, mostly concentrated along riverbanks and logging roads (Reyes-García et al. 2014). Since 1999, the Tsimane' lifestyle and knowledge have been profusely documented by a team of researchers interested in cultural change (Godoy et al. 2009; Leonard et al. 2015; Díaz-Reviriego et al. 2016). This research has highlighted the great deal of ILK across different domains maintained by the Tsimane' (e.g., Reyes-García et al. 2003; Fernández-Llamazares et al. 2015), but also the rapid erosion of Tsimane' knowledge (Reyes-García et al. 2013, 2014) largely associated to the lack of intergenerational knowledge transmission (Fernández-Llamazares et al. 2015, 2016).

In this context, over the last 15 years, a partnership of researchers working with the Tsimane' and in coordination with Tsimane' local institutions (i.e., the *Gran Consejo Tsimane'*, the legitimate political organization of the Tsimane') and community leaders has developed a number of in situ educational activities aiming at revitalizing Tsimane' knowledge and raising awareness of the multiple values of ILK. This set of initiatives has used a large range of methods and tools including (1) printed, oral, or visual educational materials on Tsimane' culture and knowledge; (2) exhibitions for the general public featuring Tsimane' culture; and (3) workshops oriented to empower Tsimane' and to raise awareness of the value of Tsimane' knowledge systems.

The partnership of researchers working in the area and Tsimane' local institutions has produced many printed, visual, and oral material featuring different aspects of Tsimane' culture. Indeed, this has been a very popular way to return research results to local communities. For example, soon after the team started research in the area, and following a request expressed by some Tsimane' and the

Gran Consejo Tsimane', researchers secured funding to elaborate a book on Tsimane' ethnobotany. The idea was discussed in meetings with a Tsimane' community and the execution counted with the participation of all the community. This initiative resulted in a book led by a Tsimane' researcher (Nate et al. 2001), which was distributed in all Tsimane' schools as an educational material. Other printed materials that have been produced as a result of this partnership include posters on uses of plants, edible fruits, and seasonal calendars. In 2003, the partnership also produced a video on Tsimane' fire making which continues to be extremely popular in local communities. In 2013, this partnership led the recording of a local radio program in Tsimane' language aimed at revitalizing different aspects of Tsimane' culture, including ILK. The program received good feedback, with some people even claiming that it had helped to "rescue from oblivion" several songs that had not been heard for years in the area (Reyes-García and Fernández-Llamazares 2019).

This partnership has also conducted exhibitions of Tsimane' culture for the wider public. An exhibition composed of the main Tsimane' handicrafts and photograph posters showing the Tsimane' way of life was set up during the annual town festival in 2001 and 2002. Handicrafts and photographs were collected in several communities in the Tsimane' territory and the material was organized around Tsimane' productive activities (e.g., hunting and fishing, maize and cassava beer making) and cultural expressions (e.g., bags and carrying tools, plant weavings, musical instruments). After the festival, the exhibition was taken to the Beni Biological Reserve Headquarters in San Borja, the main town in the Tsimane' area, and later moved to the National Ethnography and Folklore Museum in La Paz (i.e., the capital city of Bolivia). While the idea of the exhibition emerged from the researchers, all its contents were collaboratively convened with the local communities through Free, Prior and Informed Consent and an extensive consultation strategy with village leaders and elders. The exhibition served a purpose in publicly celebrating the biocultural heritage of the Tsimane' while also empowering them on the value of their own cultural knowledge.

Finally, over the years, the partnership has also fostered a number of workshops oriented to empower Tsimane' and raise awareness of the value of Tsimane' culture. For example, within the framework of a project oriented to analyze whether enhancing cultural empowerment contributes to the adoption of new farm technologies, between February and December 2001, the team conducted different types of workshops. These covered topics on (a) agriculture (the introduction of a leguminous cover crop that fitted with the traditional system of farming), (b) cultural empowerment (i.e., self-esteem; territorial rights), (c) marketing skills (aiming to improve the benefit that Tsimane' obtained in their economic exchanges – sale and barter – with traders), and (d) health (i.e., diarrhea prevention and treatment). Insights from the workshops were included in a short booklet with drawings and pictures with short commentaries in both Spanish and Tsimane' that was also widely distributed among Tsimane' communities.

11.5 Case 4: Policy/Legislation – The Recognition of Rio Negro Traditional Cultivation Systems as Cultural Heritage

Our fourth case study refers to the process that led to the recognition of the Rio Negro traditional cultivation systems as "intangible cultural heritage" by the Brazilian government, a policy/legislation initiative to protect these systems and the ILK associated to them. The Negro River is one of the major tributaries of the Amazon River, and the middle and upper Rio Negro (the focus of this case study) are situated in the northwestern Amazon region, stretching from the municipality of Barcelos until the triple border between Brazil, Colombia, and Venezuela. Due to its relative geographical isolation and limited accessibility, the region is still largely covered by native vegetation, composed of a mosaic of *terra firme* forests, flooded forests, as well as savannas and grasslands growing on white sand (Anderson 1996; Pires and Prance 1985; Goulding et al. 2003). The middle/upper Rio Negro is also one of the most ethnically diverse regions of Amazonia, home to at least 23 ethnic groups belonging to the linguistic families Tukano, Arawak, and Maku (Cabalzar and Ricardo 2006).

The traditional cultivation system practiced in the region is shifting cultivation, in which a relatively short cropping period is alternated with a longer fallow period that may be occasionally reopened for the establishment of new plots (Moran 1995). The most culturally and economically important crop is bitter manioc (*Manihot esculenta*), and the Rio Negro, where hundreds of landraces are cultivated, is recognized as an agrobiodiversity hotspot for the crop (Emperaire and Peroni 2007). Similar to other regions in Amazonia, several other annual and perennial species are also cultivated and/or managed either during the cultivation or the fallow phase, resulting in landscape mosaics composed of cultivation fields and managed fallows enriched with useful and domesticated plants (Balée and Gély 1989; Junqueira et al. 2010). Beyond their notable inter- and intra-specific biological diversity, the traditional cultivation systems of the Rio Negro comprise also the traditional knowledge and practices associated with planting, breeding and management techniques, tools and utensils used in cultivation and processing, food products and recipes, and the social networks through which plants and associated knowledge are shared (Fig. 11.1). However, just as in many other areas in Amazonia, in the middle and upper Rio Negro, the traditional cultivation systems and associated ILK are changing. Although the region is still little affected by large-scale deforestation, other changes such as rural-urban migration, increased access to the market economy, changing diets, and lack of intergenerational ILK transmission are reshaping local livelihoods and cultivation systems, often leading to the loss of ILK and genetic resources (Eloy and Lasmar 2012; Ricardo and Ricardo 2011; Emperaire and Eloy 2015).

The process that led to the recognition of Rio Negro agricultural systems as cultural heritage resulted from a dynamic interaction between local Indigenous associations, NGOs, and research and a governmental institution (Emperaire et al.

Fig. 11.1 The Rio Negro traditional cultivation systems. Left: one of the hundreds of landraces of manioc (*Manihot esculenta*), the most important and diverse crop cultivated in the region. Right, above: a woman roasting manioc flour, one of the several products derived from the processing of the crop. Right, below: an example of a diversified cultivation field, where annual and perennial crops are cultivated in shifting cultivation systems

2010). Since 1998, interdisciplinary research projects aiming to understand the process through which agrobiodiversity and ILK are constructed were developed in the region through a partnership between the Institut de Recherche pour le Développement (IRD), Instituto Socioambiental (ISA), and University of Campinas (UNICAMP) (Emperaire et al. 2010; Ricardo and Ricardo 2011). Since the start, the projects established close partnerships with local Indigenous organizations (particularly with the Association of Indigenous Communities of the Middle Negro River (ACIRMN), the Federation of Indigenous Organizations of the Negro River (FOIRN), and the Indigenous Association of Barcelos (ASIBA)), which evolved into a strong regional institutional network.

As a result of this articulation, and in the face of the growing threats to traditional cultivation systems, the need to develop strategies to guarantee the protection of these systems and associated ILK emerged. In 2007, the ACIRMN submitted a request to the Brazilian Institute for the Historic and Architectonic Heritage (IPHAN) for the recognition of the Rio Negro traditional cultivation system as "immaterial cultural heritage." This legal instrument, created by the Brazilian government in 2000, was designed to provide legal and policy support to the conservation of national cultural heritage, defined as the "doings, expressions, practices and their products, that refer to the history, memory and identity of a given people"

(Decree 3551/2000; IPHAN 2000). Once this process started, IPHAN coordinated a series of meetings and participatory research activities aiming to document the knowledge, the agrobiodiversity, and the practices associated with the regional cultivation systems which, together with the outcomes of the previous research projects, formed the basis for a dossier that finally resulted on the official recognition of the Rio Negro cultivation systems as cultural heritage in 2010 (Emperaire et al. 2010).

Beyond the increased visibility brought to the Rio Negro cultivation systems by their recognition as cultural heritage, this status also has practical implications. As part of the safeguarding strategy, the institutions involved in the process developed a detailed plan containing a series of initiatives to safeguard this heritage and established a permanent committee to monitor and evaluate the status of the system and the effectiveness of the safeguarding actions. The planned actions included, for example, activities to foster intergenerational knowledge exchange, the participatory definition of future research priorities, and the promotion of local markets with products coming from traditional cultivation systems (Emperaire et al. 2010).

11.6 Case 5: Research/Documentation – The Biozulua Database

The Biozulua database is considered as an example of a documentation initiative. This initiative took place simultaneously in 24 communities of the Venezuelan Amazonas State, an area with high biocultural diversity hosting about 75% of the country's plant species and most of the country's Indigenous ethnic groups (Zent and Zent 2007). Despite the socioeconomic and demographic changes these Indigenous populations have been experiencing in the past decades, some of the Indigenous populations still live in small communities with subsistence livelihoods based on shifting cultivation and foraging, activities that are rooted in the community's traditional knowledge. For instance, the Piaroa people of the Middle-Orinoco grow hundreds of landraces of manioc, and the Hoti people of the Sierra Maigualida know at least 220 wild edible species and 180 medicinal plants (Zent and Zent 2004, 2007; Heckler and Zent 2008). However, as in other case studies presented above, the knowledge system in the area is threatened by a continuous erosion process derived from the integration into the market economy and the adoption of the Western medicinal system, the lack of intergenerational knowledge transmission, and the exclusion of the communities from the biodiversity management plans (Zent and Zent 2007). Moreover, as in many cases around the world, the unrightfully private appropriation of this knowledge by pharmaceutical and agri-food corporations is also threatening its conservation (Vivas Eugui and Ruiz Muller 2001; Poorna et al. 2014).

In response to these threats, in the past decades, international mandates to which Venezuela subscribed have been pushing for further integrating ILK holders in environmental conservation programs and for protecting this knowledge and encouraging the sharing of any benefits associated to it (Popova 2014; Sanghera et al. 2016). In this context, in 1998 some researchers from the Venezuela Sciences Academy, who were also part of the FUDECI scientific NGO, started a project called Biozulua (Royero 2001). The project aimed at collecting knowledge in Indigenous communities via ethnobiological prospections (i.e., interviews and field visits) and storing this knowledge in a database. According to the project leaders' statements, this database would follow international mandates and protect the ILK by (1) demonstrating its existence and (2) claiming benefit redistribution to those multinational corporations interested in using it (Royero 2001; Vivas Eugui and Ruiz Muller 2001; Johnson 2002). The database collected multimedia files including descriptions of plants and plant uses, plant photos, and videos of traditional practices related to the use of those plants. These files were searchable through a specific software interface designed for the project that facilitated searching for specific knowledge in the database (Royero 2001). Plant material was also collected and stored in national herbariums and genebanks. In total the database hosted about 20,000 data entries, and the project managed to collect approximately 3000 biological specimens (Zent and Zent 2007).

Despite their key role in providing the content of the database and several mechanisms put in place to protect this content from potential misappropriation, the low participation of the IPLCs has been criticized by several authors (Zent and Zent 2007; Ochoa 2009). Criticisms include (1) the lack of participation of community members or representatives in initial stages of the project, (2) the fact that the software and structure of the database were under private property, (3) the lack of access to the database from the communities, and (4) the lack of a transparent and inclusive process to obtain Free, Prior and Informed Consent. Moreover, concerns about this projects' capacity to keep confidentiality and to empower the communities have arisen throughout the years (Long 2011; Mattie 2007). Even though the communities have property rights over the individual contents of the database, the lack of a proper regulatory framework for property rights over data compilations could imply a lack of effective property rights protection. Furthermore, the fact that the researchers act as intermediaries between interested corporations and the communities has risen alarms about the real ability of indigenous peoples to control this information. Finally, there is an overarching concern related to the de-contextualization of this knowledge, related also to the controversy over using ILK for the scientific and industrial development of the global north (Mattie 2007).

11.7 Discussion

The five ILK conservation initiatives described in this chapter illustrate some different approaches envisioned to maintain, revitalize, protect, and/or document ILK across the Amazon, one of the most important biocultural hotspots in the planet. Although we do not intend to provide an exhaustive account of the breadth and depth of all initiatives promoting participatory biocultural conservation in this region, the examples we presented indeed illustrate that there are many strategies that can help foster IPLC engagement in ILK conservation. In the following paragraphs, we will critically reflect on some factors enabling participation in these cases and draw some lessons from them.

First, the main finding that arises from this work is that different types of ILK conservation initiatives can be participatory, at least to certain degree. In other words, although the scholarly literature has shown that there is a participation gap in most ILK conservation initiatives (e.g., Benyei et al. 2019), our case studies illustrate that this participation gap can be filled. Given that ILK conservation initiatives initiated, led, and/or managed by IPLCs are potentially more legitimate than externally controlled ones (Fernández-Llamazares and Cabeza 2017), devising mechanisms to ensure IPLC participation in all phases of the ILK conservation initiatives is often a critical factor determining their success. IPLC participation requires the establishment of multi-stakeholder collaborative partnerships, which in turn requires horizontal decision-making processes that enable all voices, and in particular those of IPLCs, to be heard. In this regard, some of the initiatives described in this chapter started by (1) identifying common interests between IPLCs and other stakeholders (e.g., researchers, NGOs); (2) negotiating co-research agreements; and/or (3) outlining a mutually agreed-upon working agenda (Pert et al. 2014; Fernández-Llamazares and Cabeza 2017). In general, ILK conservation approaches are most successful when articulated from the bottom-up and/or with a strong participatory component (Packer et al. 2007; Singh et al. 2010; Gavin et al. 2015; Ryan 2015).

However, as shown in this chapter, even in those cases where the initial goals of the initiatives do not directly emerge from the local communities themselves, a number of tools and methods can be proposed to promote different levels of collaboration, participation, dialogue, co-management, and/or power sharing around these initiatives. Bringing together IPLCs, NGOs, researchers, practitioners, and governmental authorities through inclusive strategies can help to reduce the power asymmetries that have often hampered IPLC participation in ILK conservation initiatives. Yet, inclusivity requires paving a process that is considered legitimate, transparent, and equitable by the IPLCs involved and whose knowledge systems are at the core of these initiatives. To move from rhetoric to practice, several frameworks have been developed in recent years to lever power across different knowledge systems and levels of governance, such as the Multiple Evidence Based approach or the Whakatane mechanism, among many others (Tengö et al. 2014; Gavin et al. 2018). Overall, these noble goals can be best served by constant efforts to recognize and value the agency of IPLCs in these processes, challenging those approaches in which IPLCs were merely viewed as recipients or passive subjects of external initiatives. Within the Convention on Biological Diversity, there are several Indigenous codes of ethical conduct to ensure full involvement of IPLCs while respecting their cultural and intellectual heritage (e.g., Akwe: Kon Guidelines and The Tkarihwaié:ri Code of Ethical Conduct; CBD 2004, 2011). Initiatives bridging across levels of governance and with horizontal decision-making structures, such as the COICA example discussed in this chapter, are essential to better engage IPLCs in ILK conservation for reasons of social justice and more inclusive governance.

Second, our five case studies suggest that participation in ILK conservation initiatives is likely to be best achieved by in situ approaches than externally based ones. In this regard, the examples shown in this analysis indicate that in situ initiatives guided by IPLC epistemologies, needs, and views have more potential to be inclusive (Singh et al. 2010; McCarter et al. 2014; Tang and Gavin 2016; Sterling et al. 2017b). Our review aligns with a growing body of literature arguing that ILK conservation should not overlook the local social-ecological context in which ILK is generated, shared, and transmitted (Agrawal 2002; Gómez-Baggethun and Reyes-García 2013; McCarter et al. 2014). While policy and legislation support to conserve ILK is needed at multiple scales (Tang and Gavin 2016), in situ approaches are crucial in leveraging policy and legislation initiatives that are suited to local contexts and demands, as exemplified in the Rio Negro case study. The documentation of the five cases examined generally emphasizes the importance of respecting the customary mechanisms of community control, ownership, and transmission of ILK and explicitly recognizing IPLC rights and institutions, as key to the success of these initiatives.

Finally, a main lesson from all the cases reviewed is that an effective IPLC participation strategy does not happen overnight and it often requires planning it with a long-term perspective. The literature has often highlighted that punctual one-off ILK conservation initiatives are likely to be less participatory than collaborative projects developed out of sustained long-term relationships and social capital built over the years (e.g., Mulrennan et al. 2012; Sterling et al. 2017a). This is so because engaging IPLC leadership and establishing partnerships with the legitimate IPLC governance structures requires building and nurturing relationships of mutual trust over time, as shown in the Tsimane' example. Similarly, in Rio Negro, policy changes emerged as a response to the demands of IPLCs, constructed with their strong involvement and catalyzed by a long-term collaboration with research institutions and NGOs. However, long-term access to financial and technical support is often needed to build this social capital, as observed in the weaving knowledge program among the Kaiabi. Capacity-building and sustained funding support are often critical conditions to ensure IPLC engagement in these projects over long periods of time.

11.8 Conclusions

The recent changes in the Amazon's political climate suggest that the role of IPLCs in conserving the world's largest standing rainforest will be more critical than ever (Begotti 2019). As a result, there is an urgent need to devise and strengthen mechanisms not only to conserve and revitalize ILK across the whole Amazon Basin but also to actively promote IPLC engagement and support their collective action in these endeavors. Our review illustrates several examples of ILK conservation initiatives offering substantial opportunities for meaningful IPLC participation over the long term. Overall, the examples selected suggest that the development of robust and inclusive decision-making processes is essential to optimize IPLC participation in ILK conservation, thereby increasing the legitimacy of these initiatives. We believe that the lessons derived from this chapter can inspire new avenues for leading the participatory turn in biocultural conservation. Continuous political and financial support for ensuring IPLC participation in ILK conservation initiatives is therefore crucial to safeguard biocultural diversity in the Amazon and elsewhere.

Acknowledgments We dedicate this work to all the Indigenous communities in the Amazon and beyond who are working to strengthen and revitalize their knowledge systems. Research leading to this work has received funding from the Academy of Finland (grant agreement nr. 311176), the Kone Foundation, the Spanish Ministry of Economy and Competitiveness (contract grant BES-2015-072155) and the European Research Council (ERC) under grant agreement No 771056-LICCI-ERC-2017-COG. This work contributes to the "María de Maeztu Unit of Excellence" (MdM-2015-0552).

References

Abah J, Mashebe P, Denuga DD (2015) Prospect of integrating African indigenous knowledge systems into the teaching of sciences in Africa. Am J Educ Res 3:668–673

Agrawal A (2002) Indigenous knowledge and the politics of classification. Int Soc Sci J 54:287–297

Aikenhead G (2001) Integrating western and aboriginal sciences: cross-cultural science teaching. Res Sci Educ 31:337–355

Alexander M, Chamundeeswari K, Kambu A et al (2004) The role of registers and databases in the protection of traditional knowledge. UNU-IAS, Tokyo

Anderson MK (1996) Tending the wilderness. Restor Manag Notes 14:154–166

Artaxo P (2019) Working together for Amazonia. Science (80-) 363:323

Aswani S, Lemahieu A, Sauer WHH (2018) Global trends of local ecological knowledge and future implications. PLoS One 13:1–19

Athayde S (2006) O livro da cestaria Kaiabi – yrupema re je mu'e. Instituto Socioambiental, Canarana

Athayde S, Schmink M (2014) "Adaptive resistance," Conservation, and development in the Brazilian Amazon: Contradictions of political organization and empowerment in the Kaiabi diaspora. Ethnohistory 61:549–574

Athayde S, Silva-Lugo J (2018) Adaptive strategies to displacement and environmental change among the Kaiabi indigenous people of the Brazilian Amazon. Soc Nat Resour 31:666–682

Athayde S, Da Silva M, Kaiabi J et al (2006) Participatory research and management of Arumâ (Ischnosiphon gracilis [Rudge Köern.], Marantaceae) by the Kaiabi people in the Brazilian Amazon. J Ethnobiol 26:36–59

Athayde SF, Kaiabi A, Ono KY, Alexiades MN (2009) Weaving power: displacement and the dynamics of basketry knowledge amongst the Kaiabi in the Brazilian Amazon. Mobil Migr Indig Amaz Contemp Ethnoecological Perspect 11:249–274

Athayde S, Silva-Lugo J, Schmink M, Heckenberger M (2017a) The same, but different: indigenous knowledge retention, erosion, and innovation in the Brazilian Amazon. Hum Ecol 45:533–544

Athayde S, Silva-Lugo J, Schmink M et al (2017b) Reconnecting art and science for sustainability: learning from indigenous knowledge through participatory action-research in the Amazon. Ecol Soc 22:36

Balée W, Gély A (1989) Managed forest succession in Amazonia: the Ka'apor case. Adv Econ Bot 7:129–158

Ban NC, Frid A, Reid M et al (2018) Incorporate indigenous perspectives for impactful research and effective management. Nat Ecol Evol 2:1680–1683
Begotti RA (2019) Brazil's indigenous lands under threat. Science 363:592
Benyei P, Arreola G, Reyes-García V (2019) Storing and sharing: a review of indigenous and local knowledge conservation initiatives. Ambio 49:218. https://doi.org/10.1007/s13280-019-01153-6
Blackman A, Corral L, Lima ES, Asner GP (2017) Titling indigenous communities protects forests in the Peruvian Amazon. Proc Natl Acad Sci 114:4123–4128
Brodt SB (1999) Interactions of formal and informal knowledge systems in village-based tree management in central India. Agric Human Values 16:355–363
Brondizio ES, Le Tourneau F-M (2016) Environmental governance for all. Science 352:1272–1273
Brosi BJ, Balick MJ, Wolkow R et al (2007) Cultural erosion and biodiversity: canoe-making knowledge in Pohnpei, Micronesia. Conserv Biol 21:875–879
Bussmann R, Paniagua-Zambrana NY, Hart RE et al (2018) Research methods leading to a perception of knowledge loss – one century of plant use documentation among the Chácobo in Bolivia. Econ Bot 72:81–93
Cabalzar A, Ricardo CA (2006) Povos Indígenas Do Rio Negro: Uma Introdução à Diversidade Socioambiental Do Noroeste Da Amazônia Brasileira. Instituto Socioambiental, São Paulo
Castagno AE, Brayboy BMJ (2008) Culturally responsive schooling for indigenous youth: a review of the literature. Rev Educ Res 78:941–993
CBD (2004) Akwé: Kon guidelines. CBD Guideline Series, Montréal
CBD (2011) The Tkarihwaié:ri code of ethical conduct. CBD Guidelines Series, Montréal
Codato D, Pappalardo SE, Diantini A et al (2019) Oil production, biodiversity conservation and indigenous territories: towards geographical criteria for unburnable carbon areas in the Amazon rainforest. Appl Geogr 102:28–38
COICA (2011) Mandato de Manaus: Acción Indígena por la Vida. COICA, Manaus
COICA (2013) Mandato Guayupés: Amazonía Indígena: Vida Plena Amazónica frente al IIRSA y Desarrollismo. COICA, Villavicencio
COICA (2016) Mandato de la III Cumbre Amazónica: Catástrofe Climática, Amazonía Viva y Alternativas Indígenas. COICA, Lima
COICA (2018) Mandato de Macapá: Amazonía Viva Humanidad Segura. COICA, Macapá
Comunicaciones COICA (2018) Jóvenes indígenas reconocen a la COICA y piden ser incluidos. COICA, Quito
Cox PA (2000) Will tribal knowledge survive the millennium? Science 287:44–45
Díaz-Reviriego I, Fernández-Llamazares Á, Salpeteur M et al (2016) Gendered medicinal plant knowledge contributions to adaptive capacity and health sovereignty in Amazonia. Ambio 45:263–275
Eloy L, Lasmar C (2012) Urbanisation and transformation of indigenous resource management: the case of upper Rio Negro (Brazil). Int J Sustain Soc 4:372–388
Emperaire L, Eloy L (2015) Amerindian agriculture in an Urbanising Amazonia (Rio Negro, Brazil). Bull Lat Am Res 34:70–84
Emperaire L, Peroni N (2007) Traditional management of agrobiodiversity in Brazil: a case study of Manioc. Hum Ecol 35:761–768
Emperaire L, Velthem LH, Oliveira AG et al (2010) Dossiê de Registro Do Sistema Agrícola Tradicional Do Rio Negro. ACIMRN, Brasilia
Escobar H (2018) Scientists, environmentalists brace for Brazil's right turn. Science (80-) 362:273–274
Ferguson MA, Messier F (1997) Collection and analysis of traditional ecological knowledge about a population of Arctic tundra caribou. Arctic 50:17–28
Fernández-Llamazares Á, Cabeza M (2017) Rediscovering the potential of indigenous storytelling for conservation practice. Conserv Lett 11:e12398

Fernández-Llamazares Á, Díaz-Reviriego I, Luz AC et al (2015) Rapid ecosystem change challenges the adaptive capacity of local environmental knowledge. Glob Environ Chang 31:272–284

Fernández-Llamazares Á, Díaz-Reviriego I, Guèze M et al (2016) Local perceptions as a guide for the sustainable management of natural resources: empirical evidence from a small-scale society in Bolivian Amazonia. Ecol Soc 21:2

Gadgil M, Seshagiri Rao PR, Utkarsh G et al (2000) New meanings for old knowledge: the people's biodiversity registers program. Ecol Appl 10:1307–1317

Garnett ST, Burgess ND, Fa JE et al (2018) A spatial overview of the global importance of indigenous lands for conservation. Nat Sustain 1:369–374

Gavin MC, McCarter J, Mead A et al (2015) Defining biocultural approaches to conservation. Trends Ecol Evol 30:140–145

Gavin MC, McCarter J, Berkes F et al (2018) Effective biodiversity conservation requires dynamic, pluralistic, partnership-based approaches. Sustainability 10:1–11

Godoy R, Reyes-García V, Byron E et al (2005) The effect of market economies on the well-being of indigenous peoples and on their use of renewable natural resources. Annu Rev Anthropol 34:121–138

Godoy R, Reyes-García V, Gravlee CC et al (2009) Moving beyond a snapshot to understand changes in the well-being of native Amazonians: panel evidence (2002-2006) from Bolivia. Curr Anthropol 50:563–572

Gómez-Baggethun E, Reyes-García V (2013) Reinterpreting change in traditional ecological knowledge. Hum Ecol 41:643–647

Gorenflo LJ, Romaine S, Mittermeier RA, Walker-Painemilla K (2012) Co-occurrence of linguistic and biological diversity in biodiversity hotspots and high biodiversity wilderness areas. Proc Natl Acad Sci 109:8032–8037

Goulding M, Barthem R, Ferreira EJG (2003) The Smithsonian atlas of the Amazon. Smithsonian Books, Washington DC

Grünberg G (2004) O contraste entre ambientes. Instituto Socioambiental, São Paulo

Guèze M, Luz AC, Paneque-Gálvez J et al (2015) Shifts in indigenous culture relate to forest tree diversity: a case study from the Tsimane', Bolivian Amazon. Biol Conserv 186:251–259

Hamlin ML (2013) "Yo soy indígena": identifying and using traditional ecological knowledge (TEK) to make the teaching of science culturally responsive for Maya girls. Cult Stud Sci Educ 8:759–776

Heckler S, Zent S (2008) Piaroa manioc varietals: hyperdiversity or social currency? Hum Ecol 36:679–697

Herrera MP (2016) Redes transnacionales de organizaciones indígenas. Análisis del uso de las redes en conflictos socioambientales. Rev Estud Soc:63–72

Hoorn C, Wesselingh FP, ter Steege H et al (2010) Amazonia through time: Andean. Science 330:927–931

IPHAN (2000) O Registro Do Patrimônio Imaterial. Dossiê Final Das Atividades Da Comissão e Do Grupo de Trabalho Patrimônio Imaterial. Instituto do Patrimônio Histórico e Artístico Nacional, Brasilia

Jacanamijoy Tisoy SA (2011) Historia de participación de la Coordinadora de las Organizaciones Indígenas de la Cuenca Amazónica en la agenda internacional del Convenio de la Diversidad Biológica, en el artículo 8j y artículos conexos. Universidad Nacional de Colombia, Bogotá

Johnson O (2002) Venezuelan project establishes indigenous plant database. BMJ 325:183

Junqueira AB, Shepard GH, Clement CR (2010) Secondary forests on anthropogenic soils in Brazilian Amazonia conserve agrobiodiversity. Biodivers Conserv 19:1933–1961

Kimmerer RW (2002) Weaving traditional ecological knowledge into biological education: a call to action. Bioscience 52:432

Klein J (2011) Indigenous knowledge and education – the case of the Nama people in Namibia. Educ as Chang 15:81–94

Le Tourneau FM (2015) The sustainability challenges of indigenous territories in Brazil's Amazonia. Curr Opin Environ Sustain 14:213–220
Leonard WR, Reyes-García V, Tanner S et al (2015) The Tsimane' Amazonian Panel Study (TAPS): nine years (2002-2010) of annual data available to the public. Econ Hum Biol 19:51–61
Little PE (2005) Indigenous peoples and sustainable development subprojects in Brazilian Amazonia: the challenges of interculturality. Law Policy 27:450–471
Loh J, Harmon D (2005) A global index of biocultural diversity. Ecol Indic 5:231–241
Long DE (2011) Trade secrets and traditional knowledge: strengthening international protection of indigenous innovation. In: Dreyfuss RC, Strandburg KJ (eds) The law and theory of trade secrecy: a handbook of contemporary research. Edward Elgar Publishing, Cheltenham, pp 1–33
López-Maldonado Y, Berkes F (2017) Restoring the environment, revitalizing the culture: cenote conservation in Yucatan, Mexico. Ecol Soc 22:7
Maffi L (2005) Linguistic, cultural, and biological diversity. Annu Rev Anthropol 34:599–617
Maikhuri RK, Rao KS, Kandari LS et al (2005) Does the outreach programme make an impact? A case study of medicinal and aromatic plant cultivation in Uttaranchal. Curr Sci 88:1480–1486
Mato D (2000) Transnational networking and the social production of representations of identities by indigenous peoples' organizations of Latin America. Int Sociol 15:343–360
Mattie M (2007) Biozulua. Conocimiento ancestral y biodiversidad en Venezuela. In: Mattie M (ed) La Economía No Deja Ver el Bosque. Libros en Red, Buenos Aires
McCarter J, Gavin MC (2011) Perceptions of the value of traditional ecological knowledge to formal school curricula: opportunities and challenges from Malekula Island, Vanuatu. J Ethnobiol Ethnomed 7:38
McCarter J, Gavin MC (2014) In situ maintenance of traditional ecological knowledge on Malekula Island, Vanuatu. Soc Nat Resour 27:1115–1129
McCarter J, Gavin MC, Baereleo S, Love M (2014) The challenges of maintaining indigenous ecological knowledge. Ecol Soc 19:39
Moran EF (1995) Rich and poor ecosystems of Amazonia : an approach to management. In: Nishizawa T, Uitto J (eds) The fragile tropics of Latin America: sustainable management of changing environments. United Nations University Press, Tokyo, pp 45–67
Mulrennan ME, Mark R, Scott CH (2012) Revamping community-based conservation through participatory research. Can Geogr 56:243–259
Nate A, Ista D, Reyes-García V (2001) Plantas Útiles y su Aprovechamiento en la Comunidad Tsimane' de Yaranda. CIDOB-DFID, Santa Cruz de la Sierra
Nolte C, Agrawal A, Silvius KM, Soares-Filho BS (2013) Governance regime and location influence avoided deforestation success of protected areas in the Brazilian Amazon. Proc Natl Acad Sci 110:4956–4961
Ochoa MJ (2009) Design and functions of databases on TK – the case of Venezuela. In: Kamau EC, Winter G (eds) Genetic resources, traditional knowledge and the law. Solutions for access and benefit sharing. Earthscan, London, pp 327–339
Poorna LR, Mymoon M, Hariharan A (2014) Preservation and protection of traditional knowledge – diverse documentation initiatives across the globe. Curr Sci 107:1240–1246
Packer L, Rankin P, Hansteen-Izora R (2007) Living cultural storybases: self-empowering narratives for minority cultures. AEN J 2:38–46
Paneque-Gálvez J, Mas JF, Guèze M et al (2013) Land tenure and forest cover change. The case of southwestern Beni, Bolivian Amazon, 1986-2009. Appl Geogr 43:113–126
Pardo-de-Santayana M (2014) Etnobotánica e Inventario Español de Conocimientos Tradicionales. Conserv Veg 18:1–4
Pert PL, Hill R, Maclean K et al (2014) Mapping cultural ecosystem services with rainforest aboriginal peoples: integrating biocultural diversity, governance and social variation. Ecosyst Serv 13:41–56
Pires JM, Prance GT (1985) The vegetation types of the Brazilian Amazon. In: Prance GT, Lovejoy TE (eds) Key environments: Amazonia. Pergamon Press, Oxford, pp 109–115

Popova U (2014) Conservation, traditional knowledge, and indigenous peoples. Am Behav Sci 58:197–214

Pretty J, Adams B, Berkes F et al (2009) The intersections of biological diversity and cultural diversity: towards integration. Conserv Soc 7:100–112

Reyes-García V, Fernández-Llamazares Á (2019) Sing to learn: the role of songs in the transmission of indigenous knowledge among the Tsimane' of Bolivian Amazonia. J Ethnobiol 39:460–477

Reyes-García V, Godoy R, Vadez V et al (2003) Ethnobotanical knowledge shared widely among Tsimane' Amerindians, Bolivia. Science 299:1707

Reyes-García V, Vadez V, Huanca T et al (2007) Economic development and local ecological knowledge: a deadlock? Quantitative research from a native Amazonian society. Hum Ecol 35:371–377

Reyes-García V, Guèze M, Luz AC et al (2013) Evidence of traditional knowledge loss among a contemporary indigenous society. Evol Hum Behav 34:249–257

Reyes-García V, Paneque-Gálvez J, Luz A et al (2014) Cultural change and traditional ecological knowledge: an empirical analysis from the Tsimane' in the Bolivian Amazon. Hum Organ 73:162–173

Reyes-García V, Fernández-Llamazares Á, McElwee P et al (2018) The contributions of indigenous peoples and local communities to ecological restoration. Restor Ecol 27:3–8

Ricardo B, Ricardo F (2011) Povos Indígenas No Brasil 2006/2010. Instituto Socioambiental, São Paulo

Royero R (2001) Seminario Nacional de la OMPI sobre Propiedad Intelectual, Conocimientos Tradicionales y Recursos Genéticos. In: OMPI (ed) Experiencia en la Subregión Andina en materia de Documentación de los Conocimientos Tradicionales y de los Recursos Genéticos: La Experiencia Venezolana. OMPI, Quito

Rozzi R (2012) Biocultural ethics: recovering the vital links between the inhabitants, their habits, and habitats. Environ Ethics 34:27–50

Ruiz-Mallén I, Barraza L, Bodenhorn B et al (2010) Contextualising learning through the participatory construction of an environmental education programme. Int J Sci Educ 32:1755–1770

Ryan JC (2015) The virtual and the vegetal: creating a 'living' biocultural heritage archive through digital storytelling approaches. Glob Media J 9

Sanghera GS, Bhatia D, Thind KS (2016) Access and benefit sharing on the use of indigenous traditional knowledge. In: Salgotra RK, Gupta BB (eds) Plant genetic resources and traditional knowledge for food security. Springer, Singapore, pp 163–181

Schleicher J, Peres CA, Amano T et al (2017) Conservation performance of different conservation governance regimes in the Peruvian Amazon. Sci Rep 7:11318

Simpson BS, Claudie DJ, Smith NM et al (2013) Learning from both sides: experiences and opportunities in the investigation of Australian aboriginal medicinal plants. J Pharm Sci 16:259–271

Singh RK, Pretty J, Pilgrim S (2010) Traditional knowledge and biocultural diversity: learning from tribal communities for sustainable development in Northeast India. J Environ Plan Manag 53:511–533

Sterling EJ, Betley E, Sigouin A et al (2017a) Assessing the evidence for stakeholder engagement in biodiversity conservation. Biol Conserv 209:159–171

Sterling EJ, Filardi C, Toomey A et al (2017b) Biocultural approaches to well-being and sustainability indicators across scales. Nat Ecol Evol 1:1798–1806

Subba Rao S (2006) Indigenous knowledge organization: an Indian scenario. Int J Inf Manag 26:224–233

Tang R, Gavin MC (2016) A classification of threats to traditional ecological knowledge and conservation responses. Conserv Soc 14:57–70

Tengö M, Brondizio ES, Elmqvist T et al (2014) Connecting diverse knowledge systems for enhanced ecosystem governance: the multiple evidence base approach. Ambio 43:579–591

Turner NJ, Turner KL (2008) "Where our women used to get the food": cumulative effects and loss of ethnobotanical knowledge and practice; case study from coastal British Columbia. Botany 115:103–115

Varese S (1995) Pueblos indígenas y globalización en el umbral del tercer milenio. In: Barabas A, Bartolomé M, Nahmad S (eds) Articulación de la diversidad: pluralidad étnica, autonomías y democratización en América Latina. Ediciones Abya-Yala, Quito, p 410

Vivas Eugui D, Ruiz Muller M (2001) Handboook on mechanisms to protect the traditional knowledge of the Andean region indigenous communities. UNCTAD, Geneva

Zent EL, Zent S (2004) Floristic composition, structure, and diversity of four forest plots in the Sierra Maigualida, Venezuelan Guayana. Biodivers Conserv 13:2453–2483

Zent S, Zent EL (2007) On biocultural diversity from a Venezuelan perspective: tracing the interrelationships among biodiversity, culture change and legal reforms. In: McManis CR (ed) Biodiversity and the law: intellectual property. Biotechnology and Traditional Knowledge. Earthscan, London, pp 91–114

Chapter 12
Avoiding Deforestation and the Environmentalism of the Poor

Jesús García Latorre

> *Our Mother Earth, militarized, fenced-in, poisoned, a place where basic rights are systematically violated, demands that we take action. Berta Cáceres (1971–2016)*

12.1 Introduction

Tropical forest ecosystems provide manifold environmental, socio-cultural, and economic resources (e.g., Montagnini and Jordan 2005). Some of these services, like climate regulation and carbon sequestration, go beyond the local level; indeed, they benefit the whole planet. Nonetheless, humans invest tremendous efforts spoiling forests in developing countries on a large-scale. In this regard, an area encompassing no less than 179,000 km^2 has been deforested since the 1980s (Song et al. 2018). Recently, between 2010 and 2015, these ecosystems were lost at a rate of 5.5 million hectares annually (Keenan et al. 2015).

Deforestation accounts for around 10% of global greenhouse gas emissions (Angelsen et al. 2018a). Consequently, this issue has been addressed in the United Nations Framework Convention on Climate Change (UNFCCC), where the initiative Reducing Emissions from Deforestation and Forest Degradation (REDD+) has emerged. Currently, this is the most important framework for the implementation of policies to avoid deforestation in developing countries. REDD+ has triggered considerable interest in international forest-related cooperation and mobilized substantial funding (Angelsen et al. 2018b). Despite this unprecedented success, this chapter points

This paper represents the opinions of the author and does not reflect the position of his institution.

J. García Latorre (✉)
Federal Ministry for Climate Action, Environment, Energy, Mobility, Innovation and Technology, Vienna, Austria
e-mail: jesus.garcia-latorre@bmk.gv.at

C. Baldauf (ed.), *Participatory Biodiversity Conservation*,
https://doi.org/10.1007/978-3-030-41686-7_12

particularly to the technocratic approach favored in the implementation of REDD+ that limits greatly further achievements. In contrast, since the relationships between societies and the environment are shaped by socioeconomic conditions and power relations, both should be taken into account in the formulation of policies to avoid deforestation. From this standpoint, a priority role inhibiting deforestation corresponds to local and indigenous communities, not only in the framework of the implementation of policies to avoid deforestation but also through their protests and resistance.

Agribusiness is a main deforestation driver (Song et al. 2018), though other extractive industries like mining and industrial logging also play a role. These activities are carried out by private enterprises supported by the states and cause serious negative impacts on local communities and their resources (Martínez Alier 2011; Brand and Wissen 2017). Forest ecosystems represent a considerable source of environmental income, particularly for poor people (Vedeld et al. 2007). Indeed, around 1.3 billion people in the world depend, to varying degrees, on forests for their livelihoods (Forest People Programme 2012). Examples in this chapter illustrate how numerous local communities in the Global South sustainably manage their resources. These communities disproportionately feel the negative externalities of business activities and frequently complain and resist. They are the protagonists of *the environmentalism of the poor* (Martínez Alier 2011) and could play a prominent role in inhibiting deforestation if they had a say.

This chapter further addresses three issues of relevance to strengthen local and indigenous communities and to implement policies to avoid deforestation: firstly, the differential effects in a country of extractivist and inclusive institutions; secondly, new approaches for the forest-related cooperation; and, thirdly, political struggles seeking fundamental changes. The expansion of agribusinesses, deforestation, and the marginalization of rural communities exemplify the dominance of extractive political and economic institutions (Acemoglu and Robinson 2012) in developing countries (Martínez Alier 2011; Brand and Wissen 2017). A shift toward a more inclusive institutional frame is, correspondingly, a prerequisite not only to facilitate the effective involvement of local communities in the implementation of policies to avoid deforestation but also for the success of the claims and fights of indigenous peoples and subsistence peasants against extractivism. In this regard, international cooperation should focus on direct support to rural communities and aim at strengthening those communities. Finally, taking into account the prevalent hostile attitude toward rural subsistence communities, the role of nonviolent political struggles in pressuring the state to bring about fundamental changes (Petras and Veltmeyer 2007) that contribute to avoid deforestation and, in general, to sustainability is also discussed.

12.2 Policies to Avoid Deforestation

In December 2005, a group of developing countries proposed to address Reducing Emissions from Deforestation (RED) under the UNFCCC. Through the years, the original idea of RED developed into the framework "Reducing Emissions from

Deforestation and Forest Degradation and the role of conservation, sustainable management of forests and enhancement of forest carbon stocks in developing countries" (REDD+). This detailed denomination already encompasses the intended activities to tackle the problem. Table 12.1 lists the tasks that developing countries should fulfill to implement REDD+. The goal is that fully measurable, reportable, and verifiable activities lead to results-based actions, and these, in turn, lead to results-based payments (UNFCCC decisions 1/CP.16 and 9/CP.19).

Scholars have raised concerns, particularly in relation to the link of REDD+ to a market mechanism in the future (McAfee 2012; Fletcher et al. 2016) as well as to the potential negative impacts of the implementation of policies to avoid deforestation on local communities (Griffiths 2007). It should be also noted that the technocratic approach featured by the initiative and favored in the implementation seriously compromises its success. Certainly, REDD+ comprises policy and governance topics (see Table 12.1). However, more weight is put on technical issues. Countries should develop national strategies and provide information on the implementation of safeguards. Nevertheless, only the forest reference level and the information provided by the national monitoring system are subject to a technical assessment. The forest reference level represents a quantification of expected future emissions that are compared with actual emissions at the end of a commitment period (Dooley and Gupta 2017). This allows one to illustrate whether emissions from deforestation in a country have decreased. However, this approach has been contested because it "open[s] up the possibility of inflating future expectations of emissions, in order to make targets easier to meet" (Dooley and Gupta 2017). That was the case for the first reference level of REDD+ that was presented by a country, Brazil. The authors that reviewed it concluded that such constructions "show a tendency towards profitable national reference level definitions with the risk of undermining climate effectiveness and cost-efficiency of the REDD+ mechanism" (Hargita et al. 2016). Furthermore, technocratic bias is also confirmed by the fact that countries are placing particular emphasis on forest monitoring and finance aspects while considerably neglecting forest governance and the implementation of safeguards (Thu Thuy et al. 2018).

The importance of monitoring systems should by no means be undervalued. Nonetheless, technology alone is not sufficient. The development of REDD+ at the national level during recent years highlights the difficulties faced by the implementation of policies to avoid deforestation. Powerful actors, e.g., private companies,

Table 12.1 An overview of the tasks that developing countries should fulfill when implementing Reducing Emissions from Deforestation and Forest Degradation (REDD+). Activities with an asterisk are subject to a technical assessment. Sources: decisions 1/CP. 16 (2010) and 4/CP.15 (2009) of the UNFCCC

Required activities for the implementation of REDD+
National strategy or action plan (It should address deforestation drivers, land-tenure issues, gender issues, and safeguards)
Forest reference level*
National forest monitoring system*
Summary of information regarding how safeguards are being addressed and respected

benefiting from continued forest conversion, resist national policy reforms and block them (Angelsen et al. 2018c). Since political commitment is highly influenced by economic policy (Thu Thuy et al. 2018), those deforesting have been successful in achieving their goals (Angelsen et al. 2018a). Deforestation drivers, like export-oriented agriculture and weak forest governance, are not widely recognized by governments, and the required policy reforms are currently missing (Thu Thuy et al. 2018). This economic-dominated context leads to the astounding situation in which efforts to implement REDD+ coincide with continued support for development programs for large-scale conversion (Thu Thuy et al. 2018). Thus, for instance, the government's command-and-control measures to tackle deforestation in Brazil were supported crucially by the use of Landsat data on forest cover and loss and led ultimately to an important decline in deforestation after 2004 (Hansen et al. 2013; Nepstad et al. 2014). However, this approach has neither changed the frame of the extractive institutions in the country (Amazon Watch 2018) nor has it impeded the recent worsening of deforestation (Weisse and Ruiz 2019).

Thus, the issue of deforestation is not principally technological but political and socio-economic, just like the reforms required to tackle it. The next section treats, in depth, this assertion and sheds light on the factors influencing how humans deal with natural resources.

12.3 Societies and the Environment

An understanding "of the combined system of humans and nature is needed to formulate policies" (Holling et al. 1998). Thus, a short incursion into the relationships between societies and forests would help facilitate the implementation of ways to improve policies to avoid deforestation.

Research carried out in the fields of environmental history and anthropology shows how the choices that a society makes respecting the use of its natural resources reflect its socioeconomic structure and relations of power (Godelier 1981; Worster 1988; Radkau 2012). Several examples from Europe should clarify this perspective.

Until the end of the fifteenth century, a Medieval Islamic kingdom persisted in the southeast of the Iberian Peninsula, the Nasrid Kingdom of Granada. At present, semi-arid landscapes dominate this territory. However, at that time, trees and forests played an important role in the area. This was the consequence of a particular socioeconomic organization. In the rest of the Iberian Peninsula, a European feudal society developed, featuring two main elements: peasant communities and a warrior class. The former were obligated to grow a surplus of crops, especially cereals that could be stored long term. The warriors, for their part, controlled the peasant class and claimed their surplus. In contrast, the Muslim society exhibited a more egalitarian social structure. A free peasantry produced, for itself, mostly perishable crops that could not be easily stored and hence could not serve as the foundation of a feudal system. They favored crops like fruit trees and vegetables and concentrated their activities in tiny spaces in the bottoms of ravines and valleys where irrigation was possible. The rest of the territory, unlike the area-demanding cereal agriculture of feudal Europe, was used in an extensive way. As a result, deforestation was limited, and forests existed even under semi-arid conditions (García Latorre et al. 2001).

On the other hand, population growth and agricultural innovations within the feudal structure in Medieval Central Europe led to "one of the great deforestation episodes in the world" (Williams 2003), replacing half of the forest area by cereal fields. These two contrasting cases exemplify the environmental consequences of two different socioeconomic forms.

Villages reacted to increasing conflicts around the shrinking forests by organizing themselves to collectively use forest commons (Meiners and Rösener 2004). These commons provided, in the frame of a subsistence economy, manifold resources and played an important role for landless poor people. Complex sets of norms regulated their use. Nevertheless, in preindustrial times, wood was the main source of energy for relevant activities like mines and ironworks. Thus, and in contrast to the diversified management of forests by the peasantry, local lords and sovereigns favored the economic value of timber. The ruling class accused villagers of degrading the forests and implemented forest protection by-laws as an instrument of power (Radkau 2012). Again, two conflicting socioeconomic backgrounds sought to apply management approaches leading to different impacts on the forests. While local communities pursued the conservation of woodlands for the provision of manifold resources, the elite exerted power to establish monocultures (Radkau 2012).

It was in this conflicting context that scientific forestry emerged first in Germany, before spreading later to the rest of the world (Scott 1998). Forestry developed as an aspect of state administration serving the elite's economic interests. It focused on the quantification of timber and the "rationalization" of the management of forests. This approach led to monocultural even-aged stands, and foresters perceived geometric perfection "as the outward sign of the well-managed forest" (Scott 1998). As expected, this new forest management clashed with local communities, whose practices were regarded by foresters as a threat to the forest. Nowadays, forestry practitioners still overlook the socioeconomic context of the origin of their discipline. They interpret the emergence of forestry as a mere technical response to wood scarcity not linked to political interest and treat forestry as only a technical field (García Latorre and García Latorre 2012).

Against this background, Fig. 12.1 depicts the framework of the societal management of natural resources. The described cases illuminate how the socioeconomic form of a particular group is a determinant for its management of forests. Furthermore, the techno-environment refers to the application of technology, like the irrigated crops of Medieval Muslims contrasted with the area-demanding cereal fields of Christian peasants, or the manifold resources managed in the communal woodlands, which differ from the monocultural forest stands. Moreover, it is power that enables one to implement a specific approach. Finally, every group constructs an own representation of the environment through a set of intangible elements like ideas, value, and myths that reflect the socioeconomic background leading to concrete management practices. Such constructions are used by societies in defining what a resource is or assessing the condition of the environment (Worster 1988), e.g., whether a forest is degraded or, conversely, well managed.

Fig. 12.1 Framework describing the societal management of natural resources. (Based on Godelier (1981), Worster (1988), Berkes and Folke (1998) and Radkau (2012))

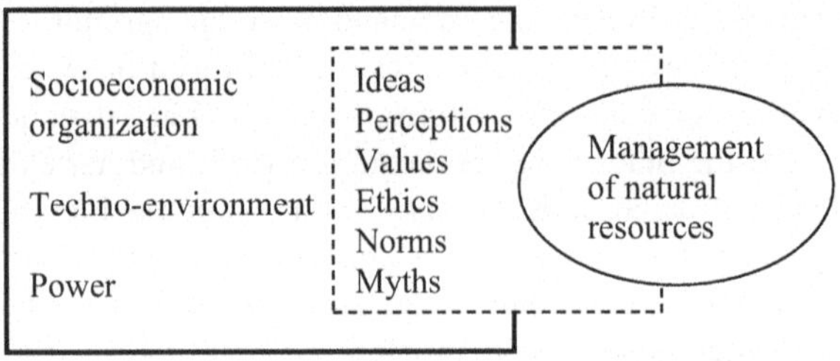

Consequently, policies to avoid deforestation that emphasize technical approaches overlook the complex socioeconomic and cultural components located at the heart of activities like the management of forests. To this effect, the next section illuminates the relevance of social groups within a country whose management practices support sustainability and avoiding deforestation.

12.4 The Environmentalism of the Poor

According to the well-known report *Our common future*, "poverty is a major cause and effect of global environmental problems" (World Commission on Environment and Development 1987). It was this publication that contributed to the wide dissemination of the conviction that poverty inescapably leads to environmental degradation (Martínez Alier 1994). In fact, this is an unfortunate prospect that prevents one from recognizing the important contribution of poor local and indigenous communities to true sustainability. Contrary to *Our common future* though, Martínez Alier (1994) emphasizes that the poor definitely support sustainability through the conservation of agricultural genetic diversity and their disproportionately low emission of greenhouse gasses. Furthermore, an analysis of the drivers of land-cover changes confirms that poverty alone does not constitute a major underlying cause of deforestation (Lambin et al. 2001).

Ironically, it is in the wealthy societies of the global north and economies in transition where patterns of production and consumption lead to appalling social and environmental impacts on the countries of the Global South. Economic growth in rich countries pushes the frontier for the extraction of natural resources to new territories (Martínez Alier 2011), underpinning the expansion of agro-businesses, industrial forestry, and mining (Global Witness 2009; Martínez Alier 2011; Guereña and Rojas Villagra 2016; Higgonet et al. 2016; Amazon Watch 2018). Recently, two terms have emerged that illuminate the socioeconomic and political context of this robber economy. Brand and Wissen (2017) underline how the *imperial mode of living*, found in well-off countries, exacerbates the impacts of climate change and leads to the destruction of ecosystems and the impoverishment of the population in developing countries. Lessenich (2017), for his part, refers to the *externalization societies* that externalize the immense costs of economic activities. Consequently, he further accentuates that the prosperity of these societies corresponds with the evils in poor countries. These negative impacts are frequently "felt disproportionately by some social groups that often complain and resist" (Martínez Alier 2011). These groups are the protagonists of the environmentalism of the poor.

Poor rural people have a material interest in the resources and services provided by the environment for their livelihoods (Martínez Alier 2011) because they directly rely on them (Davey 2009). Alluding to Fig. 12.1, this dependency stands at the core of the management systems applied by indigenous and subsistence peasant communities. Mr. Priichaa Siri, a member of the Pgaz K'Nyau (Karen) community in northern Thailand, gets to the heart of the matter by noting that "the people know that we depend on nature for our living and that if we use nature without taking care of it, or if we destroy it, then that would be the same as killing ourselves. For sure, we would never do that to ourselves" (in Trakansuphakorn and Kampholkul 2010). Hence, these groups favor a healthy environment and sustainability (Martínez Alier 2011). In this regard, women play a prominent role in environmentalism in poor rural areas. Due to the traditional gender division of labor, many activities related to the gathering of natural resources like water, firewood, vegetables, etc. fall on females. That is why women are particularly aware of the importance of these resources and easily notice when they decrease (Agarwal 2000; Davey 2009; Martínez Alier 2011): "For us", says a Miskito woman (Nicaragua), "the forest is a source of life, it is our heritage for our descendants and for this reason we cherish and protect it" (in Fenly 2011). The essence of the environmentalism of the poor is also clearly expressed by a woman from Ecuador complaining about the impacts of industrial shrimp farming on mangrove forests: "We are struggling for something which is ours, our ecosystem, but not because we are professional ecologists but because we must remain alive, because if the mangroves disappear [...] we all disappear" (in Martínez Alier 2011).

It is worth referring to the forms in which poor rural people express their representation of the environment and their concerns beyond a mere monetary valuation, emphasizing, for instance, human and territorial rights and sacredness (Martínez Alier 2011). As the indigenous peoples of Northern Mindanao (the Philippines) said during a conference, "without this ancestral land, we will not exist," clarifying that, "without the forests, mountains, rivers and our farms, we cannot continue to practice our culture" (in Guiang et al. 2001). Likewise, Mr. Makling Dulang, a Kalinga chieftain who led the struggle against the dam construction across the Chico River in the Philippines and was murdered in 1980, stated that "land is sacred and land is life" (in Guiang et al. 2001). These specific languages that appeal to non-economic values might hinder the appreciation of poor peoples' claims by politicians unable or unwilling "to understand messages encoded in terms other than those of the dominant economic discourse" (Roy Rappaport 1993, in Martínez Alier 2011).

Of particular relevance is the magnitude of the environmentalism of the poor. Hundreds of cases can be found in the Global South (Martínez Alier 2011). Some examples from nine countries illustrating how indigenous and local communities contribute to sustainability are shown in Table 12.2. These cases alone cover an area of over 24 million ha. In Brazil, indigenous communities have successfully inhibited deforestation. Their lands comprise one-fifth of the Brazilian Amazon and are

Table 12.2 Some examples of the environmentalism of the poor in forestlands

Country	Communities supporting sustainability in forestlands
The Philippines	Around four million people in local and indigenous communities manage, sustainably, six million ha of community forests. Their practices contribute to the conservation of natural resources and biodiversity. Moreover, income generation and resource rights have improved. The last intact forests are located in indigenous territories. Overall, communities in forestlands (around 24 million people) are major players in the sustainable management of forest resources
India	Local communities manage 17.33 million ha (Joint Forest Management Programme) contributing to the regeneration of degraded land and the promotion of biodiversity. Local people, unlike the forest department, favor autochthonous tree species for afforestation
Bangladesh	Around 700–800 village common forests in the Chittagong Hill Tracts, averaging from 20 to 120 ha in size, are managed by indigenous communities. Forest commons represent a traditional approach and a manifestation of rural communities' responses to forest degradation; they also play an important role in conserving natural resources and provide a model for forest restoration across Bangladesh
Thailand	A network of eight villages in the North of Thailand, the Luuk Dort River Basin Network, applies a traditional approach managing, sustainably, 9600 ha of community forests. They have successfully fought against state interventionism and hostility, as well as a forest concession in their territory
Kenia	The Naimina Enkiyio Forest (330 km^2) in the Maasai community of Loita in Southern Kenya is one of the few non-gazetted and well-conserved indigenous forests in the country. Women, particularly, contribute to the sustainable management of communal resources and oppose their privatization
Paraguay	Indigenous and subsistence peasant communities carry out agroecological practices and advocate for the conservation of forests. They protest against and oppose the expansion of the agro-business that is supported by the government and carried out, frequently, by farmers of Brazilian origin
Guatemala	In the North of Guatemala, 24 member communities organized in the Association of Forest Communities of Petén (ACOFOP) successfully manage over 426,000 ha of common forestland. This is the largest community forest concession in the world that is FSC certified. The communities have established local governance systems and have succeeded in fighting illegal logging and deforestation
Colombia	In Tumaco, African-American communities sustainably use the resources provided by the mangrove forests. These resources include shrimp, crayfish, fish, and charcoal production. The traditional gathering of shrimp and other resources represents an important source of food and cash income for women. Nevertheless, mangroves are threatened by the activities of industrial shrimp farming. These are state forests under traditional usage by local communities. Correspondingly, establishing concessions for shrimp growers constitutes a privatization perceived locally as an environmental and social "tragedy of enclosures."

(continued)

Table 12.2 (continued)

Country	Communities supporting sustainability in forestlands
Mexico	Rodolfo Montiel, a local subsistence peasant in the Sierra de Petatlán, established the Organization of Environmental Peasants. For 7 years, Rodolfo and other community members opposed the rampant logging activities carried out in the area by the forestry firm Boise Cascade. In May 1999, Rodolfo and Teodoro Cabrera were arrested by the Mexican army without an injunction from a court of law. They were tortured and put in prison for two and a half years. The Association managed to remove Boise Cascade, and Rodolfo received the Goldman Prize for the year 2000.

Sources: Guiang et al. (2001), Lasco and Pulhin (2006) for the Philippines; Murali (2002), Ravindranath et al. (2006) for India; Jashimuddin and Inoue (2012) for Bangladesh; Trakansuphakorn and Kampholkul (2010) for Thailand; Kimaren and Riamit (2011), Kariuki et al. (2016) for Kenya; own observations for Paraguay; Cronkleton et al. (2010), Sergio Guzmán (personal communication) for Guatemala; Matínez Alier (2011) for Colombia and Mexico

five times the area designated as parks (Nepstad et al. 2006). In South and Southeast Asia, over 25 million ha of forests are sustainably managed by local and indigenous communities (Ravindranath et al. 2006). Communal land tenure rights contribute to the protection of natural resources, thereby setting them apart from the market. They are of particular importance for rural women, since their access to private property resources is very limited (Agarwal 2000; Davey 2009; Martínez Alier 2011). According to Pretty (2003), there are around 0.4–0.5 million groups involving some 8–15 million households, mostly in developing countries, managing common property resources. On this subject, the community-based management of forests "may be at least, if not more, effective in reducing deforestation as protected areas at the pantropical scale" (Porter-Bolland et al. 2012). Considering the high percentage of poor people worldwide, Martínez Alier (2011) concludes that the environmentalism of the poor is concerned with "the majority of humankind, those who occupy relatively little environmental space, who have managed sustainable agroforestal and agricultural systems" and whose livelihoods are threatened by business initiatives.

The environmentalism of the poor emerges from the activism of poor women and men affected by the loss of natural resources as the basis of their livelihoods. For unjust politics, their movements are adversarial against governments; likewise, they oppose extractive projects run by states or by corporations. Consequently, thousands of environmental conflicts occur annually in southern countries (Martínez Alier 2011). Unfortunately, local communities do not receive recognition for their contribution to sustainability. In fact, they are threatened and experience violence. The cases of Chico Mendes, who faced the ranchers deforesting in Acre (Brazil) and was murdered in 1980, and Ken Saro-Wiwa, who was executed in 1995 together with eight dissenters for protesting against oil extraction by the company Shell and its military backers in the Niger Delta (Nigeria) (Martínez Alier 2011), are well known. Moreover, hundreds of murders of land and environmental defenders occur every year, leaving many cases unreported. More than 1000 murders have been documented by Global Witness since 2010, having record years in 2016 and 2017, with two hundred and two hundred one cases, respectively (Global Witness 2017a, 2018). This is a spreading phenomenon, since numbers are increasing and more places are being affected. Brazil, Honduras, and Nicaragua are among the deadliest countries in the world, with the Philippines as an outstanding example in Asia (Global Witness 2017a).

The expanding frontier for the extraction of natural resources is triggering a fight for land "as mining, logging, hydro-electric and agricultural companies trample on people and the environment in their pursuit of profit" (Global Witness 2017a). The bakers of extractive activities and infrastructure projects, including governments, companies, investors, and development banks, are all complicit in this crisis (Global Witness 2017a). This abominable problem has a particularly perverse side. As mentioned above, women have a close relationship to natural resources and play a special role supporting their livelihoods and communities (Agarwal 2000). Thus, they are frequently at the forefront of rural movements (Martínez Alier 2011). That is why, in recent years, women are being especially targeted by contract killers in South America, a region in which feminicides are increasing (Cuentas 2015; Bayona 2018). So, for instance, Jakeline Romero, a Wayúu woman in the Northwest of Colombia, has denounced the environmental damage caused by the coal mine Cerrejón: "For the Wayúu people," she says, "we are paying with our lives. We are paying with our culture. We are paying with the threat of being extinguished […] Simply because we defend this small piece of land that used to give us enough to eat" (in Global Witness 2017a). Due to her activism against the abuses of powerful corporations, Jakeline and her family face threats and intimidation. In Honduras, Berta Cáceres denounced 33 death threats from the hydro-electric company Desa (Regadas 2017). Berta protested against extractivism and the construction of a dam by Desa in the sacred river of Gualcarque. She was killed in March 2016 (Global Witness 2017b). Many other examples of this desperate situation could be added. Thus, Martínez Alier (2011) challenges whether states from the south should be trusted regarding the implementation of environmental policies, "because they have armies and pollute with state firms or sponsor polluting enterprises."

Summing up, it can be concluded that the contribution of poor rural communities to inhibiting deforestation and supporting sustainability is unquestionable. Davey (2009) emphasizes that "this form of environmentalism offers much in terms of moving towards achieving a global sustainable community". Mario Palacios Panéz, for his part, perceives this kind of activism "as a principal social player directing the economy to a more fair and sustainable way" (in Martínez Alier 2011). Furthermore, the environmentalism of the poor represents for Martínez Alier (2011) potentially "the most powerful current of environmentalism," an alternative way of modernity and development. Finally, in this regard, Angelsen et al. (2018c) highlight that secure indigenous and local community rights "in many cases could be central to successful forest-based mitigation strategies." Thus, it would be beneficial to give them a stronger political stake. With this purpose in mind, the next section addresses the requirements needed to promote a shift to inclusive institutions in developing countries.

12.5 Inclusive Institutions to Avoid Deforestation

Considering the different groups of a society along a spectrum, from the exploitation to the sustainable management and conservation of forests (Grainger and Konteh 2007), the protagonists of the environmentalism of the Poor are part and parcel of protectionist groups. On the other end of the spectrum are exploitative groups aiming at agribusiness and logging timber. Due to the influence exerted by powerful economic factions on governments (Martínez Alier 2011; Brand and Wissen 2017; Lessenich 2017; Thu Thuy et al. 2018), exploitative groups prevail (Angelsen et al. 2018c). Consequently, inhibiting deforestation implies institutional changes that allow protectionist groups to increase the effectivity of their pressure on policymakers (Grainger et al. 2006). This section illustrates the dominance of exploitative groups and discusses a path supporting inclusive institutions for the implementation of policies to avoid deforestation.

Political and economic extractive institutions promote a concentration of richness and power in the hands of elites that aims at perpetuating their influence at the expense of the vast mass of people. Such elites have a stranglehold on political and economic life and keep the country in poverty (Acemoglu and Robinson 2012). Politicians, for their part, "are just too happy to extract resources or quash any type of independent economic activity that threatens themselves and the economic elites" (Acemoglu and Robinson 2012). The intensity of these institutions differs in the countries of the Global South. However, they persist despite the presence of democratic political systems (Acemoglu and Robinson 2012). This fact is highlighted by Pimple and Sethi (2005), who state that "even in democracies laws are often mere creations of an economic and social elite, aiming to preserve the status quo, or at least ensure that reforms and other progressive legislation do not rock the boat too dangerously." This is the institutional context supporting deforestation (Grainger 2004; Grainger and Malayang 2006). The following cases in South America and Asia highlight it.

Wichí indigenous peoples sustainably use forest resources in the Argentinean Chaco. However, strong ties between soy farmers and government officials have facilitated the consolidation of agribusiness, exacerbating deforestation. Over two million hectares were deforested between 1976 and 2012, with 45% of this deforestation occurring between 2000 and 2012 (Gabay and Alam 2017). Recently, the provincial government of Salta, pressured by agri-businessmen, revoked the protection of 51,768 ha (Greenpeace 2018). This shift in land use patterns causes the marginalization and expulsion of rural communities (Gabay and Alam 2017).

Paraguay provides an example of overwhelming dominance by extractive institutions, thereby allowing the uncontrolled growth of agribusiness and widespread deforestation (Fig. 12.2). In the period from 2010 to 2014, the Paraguayan Chaco featured the highest deforestation rate worldwide, reaching 260,000 ha annually, with over 1.3 million ha of forests removed in 4 years (Guereña and Rojas Villagra 2016). Horacio Cartes, the former president, signed a decree modifying the Forest Law to clear 2000 ha in the Chaco, enlarging pastures in his property. This decree was repealed by the new government in 2018 (abc color 2018). However, this case illustrates the influence of vested interests on Paraguay's politics. Similarly, Ariel

Fig. 12.2 About 300 indigenous Ava-Guaraní people camping in the Plaza de Armas near the congress in Asunción (Paraguay) to protest their deplorable situation. They were forcibly evicted from their community by Brazilian soy farmers. The community leader's brother was murdered directly in the Plaza de Armas. They were here under miserable conditions for around 6 months. (Photo: Jesús García Latorre)

Oviedo, the new environment minister, tweeted shortly after his nomination in August 2018, "we have to be serious cooperating with and being supportive to investors and industrialists," confirming the extractivist paradigm. A consequence of this hostile framework is the highest land inequality worldwide, leaving 300,000 rural families without access to land (Guereña and Rojas Villagra 2016). Many legislators are themselves large landowners or have close relationships to the landowner oligarchy (Guereña and Rojas Villagra 2016) and thus oppose any reform attempts (Riquelme 2005). Protests by rural groups are suppressed violently, like in the massacre of Marina Kue, in which 11 peasants and 6 policemen died (Kretschmer 2014). As stated by the recognized novelist Augusto Roa Bastos, "misfortune fell in love with Paraguay."

In August 2019, widespread forest fires in the Brazilian Amazon attracted worldwide media attention. The number of fires since the beginning of the year exceeded the same period in 2018 by 39% (Weisse and Ruiz 2019). The magazine Globo Rural discovered that a group of over 70 persons from the agribusiness planned to set fires in the State of Pará seeking support for president Jair Bolsonaro's prospects to ease environmental legislation (Aranha 2019). Indeed, agribusiness' interests are well represented in the Brazilian congress, which, in turn, is leading to the worst assault on the Amazon in a generation. Alone in June 2018, a forest area encompassing 1169 km^2 was lost mostly to accommodate agricultural production (Amazon Watch 2018). A group of lawmakers, known as the ruralistas, representing landowner interests and various lobbies, is pushing for anti-indigenous legislation and removing environmental barriers to facilitate deforestation (Amazon Watch 2018). Like in Paraguay, many of them are themselves landowners or have direct connections with farming and ranching groups (Amazon Watch 2018).

The last case refers to Indonesia, where indigenous communities manage an area of 11 million ha comprising diverse forest ecosystems like rubber and rattan agroforests, which contribute to biodiversity conservation and the national economy (Fay 2009). Despite this enormous natural and human capital, the government has questioned their capacity to manage forests, blaming them as deforestation agents (Ola Kleden et al. 2009). Moreover, the powerful government elite has traditionally collaborated with the private sector to promote agribusiness, industrial logging, and mining, while local communities are marginalized and criminalized (Affif 2009). Protectionist groups claim, for the recognition of their customary law, traditional management systems, and communal land tenure rights, having received a mere grudging acknowledgment so far (Ola Kleden et al. 2009).

The historical perspective applied by Acemoglu et al. (2012) to economic development and Grainger et al. (2006) to sustainability in forests emphasizes the only way out of this vexatious situation: "to force the elite to create more pluralistic institutions" (Acemoglu and Robinson 2012). This transformation is not easy, "but it is not impossible," these authors insist. An inclusive context supports distributing power broadly in society, thereby constraining, simultaneously, its arbitrary exercise and promoting equitable access to resources (Acemoglu and Robinson 2012). This is the appropriate institutional frame in which policies to avoid deforestation can succeed. How policymakers generally perceive an issue is coming under pressure from various groups (Grainger and Malayang 2006). Thus, a broad segment of society must mobilize in order to pressure policymakers to support more inclusive institutions (Grainger and Malayang 2006; Acemoglu and Robinson 2012). In this respect, democratization and pluralization play an important role in giving diverse groups the freedom "to voice their opinions and put pressures on the government and make it more willing to listen to them" (Grainger 2004).

The previous section highlighted groups in the Global South claiming social and environmental justice. Such groups play a particular role in "leading the fight against the existing regime" (Acemoglu and Robinson 2012). In this context of conflicts, Grainger et al. (2006) highlight the "balance of effective pressure" of interest groups on policymakers as the relevant criterion for policy change. Table 12.3 summarizes the factors that improve the skills needed to transmit pressures that can allow protectionist groups to force a government to switch policy. These groups might succeed independently of their strength compared to exploitative groups, provided that they become as effective as the exploitative ones in transmitting pressure (Grainger and Malayang 2006).

Table 12.3 Factors related to the skills of interest groups transmitting pressure. These factors, together with an appropriate framework for political freedom (democratization and pluralization), determine the effective strength of the pressure of an interest group on policy-makers (based in Grainger 2004; Grainger and Malayang 2006)

Factors related to the skills of interest groups transmitting pressure
Professional promotion of the cause
Reasonable access to policymakers
Public meetings and media participation lead to public awareness
Increasing availability of finance and public support
Increasing number of pressure groups and the foundation of networks and broad coalitions

Using this "dynamic policy pressure model" to explain policy changes has been tested in Thailand and the Philippines (Grainger 2004; Grainger and Malayang 2006). In both countries, a policy shift toward sustainability in forests began to develop in the 1980s. Initially, different groups, including rural initiatives and urban-based NGOs, independently protested against eucalyptus plantations, dam projects, and other environmental damages. Subsequently, these groups joined and benefited from the support of the middle class and the media. This combination of middle class and environmental campaigners constitutes a potent force, and the support of the media contributes to successfully transmitting pressure on policymakers (Grainger 2004, see Table 12.3). Banning log exports, imposing logging moratoria, and other measures supporting sustainability were the result.

Land tenure rights are an important claim from rural communities and also provide an example of a political shift toward inclusiveness. Acemoglu et al. (2012) highlight secure private ownership rights as an aspect of inclusive institutions. Nonetheless, the communal tenure of land plays a very important role in indigenous and local communities and delivers positive sustainability outcomes (Pretty 2003). Since the mid-1980s, a shift in forest tenure has led to the transfer of around 200 million ha of forestland to local communities, a reform driven by communities and social movements, but also by international donors and the state (Larson et al. 2010). However, this success is only partial. Severe and coercive forest policies counteract tenure reforms while limiting effective participation (García Latorre and García Latorre 2012). Forest peoples continue claiming efficient legal recognition of their rights by national laws (Forest Peoples Programme 2012).

The importance of networks and coalitions that coordinate the population's demands, and are well established enough not to be easily crushed by elites, has been emphasized by Acemoglu et al. (2012). In this regard, the efforts of rural people struggling for their livelihoods have frequently coalesced in the emergence of associations and networks. Such organizations ensure tenure rights and promote community benefits and increasing bargaining power (Paudel et al. 2010). Nonetheless, their integrity remains a challenge "given the constant attempts by external actors to usurp resources" (Paudel et al. 2010). Thus, notwithstanding the importance of these efforts, a shift to solid inclusive institutions assumes a greater involvement in politics to pressure the state "to bring about fundamental changes" (Petras and Veltmeyer 2007). An example from Indonesia illustrates the difficulties associated with such an involvement. The Indigenous Peoples Alliance for the Archipelago (AMAN) strengthens people's resistance to unfair polices, environmental degradation, and human rights violations. Nonetheless, given the marginalization of indigenous communities, increased political participation is required to make policymakers more accountable to communities (Ola Kleden et al. 2009). Hence, "indigenous peoples must brave the political arena" (Syahruji and Kiyu 2009). This, however, represents a challenge, since they live in rural areas far from political centers, which in turn handicaps their participation in political decision-making processes (Ola Kleden et al. 2009; Syahruji and Kiyu 2009).

The next section analyzes the forest-related cooperation and highlights ways to support and empower protectionist groups in developing countries.

12.6 International Cooperation to Support Local and Indigenous Communities

Forest-related assistance shares a similar fate with the broader development cooperation. Despite considerable investment, international efforts have not succeeded in improving the economy of poor countries (Easterly 2006; Acemoglu and Robinson 2012). Likewise, assistance given to forestry has not halted deforestation (Persson 2000; Rukunuddin 2003; Grainger and Malayang 2006). In this regard, and considering the ideas of the previous section, it is important to emphasize that foreign aid (Acemoglu and Robinson 2012) and overseas pressures (Grainger 2004, Grainger and Malayang 2006) have only a very limited influence on the shift to inclusive institutions, the appropriate frame for sustainability. On the contrary, the most effective sources of pressures are groups based in the country (Grainger 2004). This section analyzes factors that hinder efficient cooperation and highlights alternatives to improve it.

Cooperation is mostly designed at the governmental level (Easterly 2006; Aurenhammer 2013), which allows influential stakeholders from both sides, the donor and the recipient country, to privilege their own perceptions of problems and solutions (Persson 2000; Rukunuddin 2003; Easterly 2006; Aurenhammer 2013). Certainly, donors frequently do not have an adequate understanding of the situation in the developing country (Persson 2000). However, it should not be assumed that recipient governments have a better understanding, as exemplified by the following cases. Hence, a network of eight villages in the North of Thailand successfully manages over 9000 ha of community forests by applying a traditional system of rotational farming (see Table 12.3). The sustainability of this approach is beyond question. However, "the Thai government has lacked understanding of the ways of life of indigenous peoples" and "still refuses to accept the proof" (Trakansuphakorn and Kampholkul 2010). State officials perceive this traditional management as backward and apply control and suppression against subsistence peasants. One reason for this is that the simplifying management of the German model of "scientific forestry" has become standard throughout the world (Scott 1998), and forestry practitioners have difficulties in understanding forests that are managed traditionally (García Latorre and García Latorre 2012). Thus, official agencies in Indonesia classify areas covered by agroforests encompassing crops, secondary forests, and bush fallow as "non-productive dry lands" (Syahruji and Kiyu 2009). A final example refers to West African governments that obsessively blame local communities for degrading forests through firewood gathering (author's own observation), a view broadly contested by scientists (Benjaminsen 1997). Therefore, it is not surprising that Easterly (2006) questions implementing projects in close consultation with the recipient government. Regarding donors, cooperation projects are frequently donor-driven, reflect their political and economic interests (as well as their priorities), and rely on consultant firms, which increases their costs enormously (Persson 2000; Rukunuddin 2003; Aurenhammer 2013). Correspondingly, most aid money returns to donor countries, while only small amounts go to projects that benefit the poorest people (Rukunuddin 2003). As Persson (2000) states, "altruism seems often to be declining in importance." Indeed, biased and interest-driven cooperation projects hardly address reality and will change it for the better.

A fundamental aspect of cooperation initiatives refers to the knowledge upon which they are based. Concerning this knowledge, the sense of superiority of the countries of the global North (Brand and Wissen 2017; Lessenich 2017) positioning their views and capacities solidly on projects (Aurenhammer 2013) further undercuts efficiency. Thus, forest-related cooperation frequently focuses on forestry instruments for planning and inventory and neglects traditional ecological knowledge, local capacities, and institutions (Aurenhammer 2013). Certainly, giving predominance to our knowledge "is an assumption that 'they' are ignorant" (Crewe and Harrison 1998). "It is very difficult," lament Crewe and Harrison (1998), "to shake off the idea that we know more than them and accept that we might even learn from them." This underestimation of local knowledge is aggravated by the positive recognition of recipient governments for outside know-how and capacities (Rukunuddin 2003, Aurenhammer 2013). Therefore, this context leads to a loss of approaches and instruments to solve problems (Aurenhammer 2013). It is also illustrative to consider the economic-related aspects of knowledge transfer beyond the benefits for donors' consultants. Martínez Alier (2011) refers to the coincidence of replacing traditional practices by foreign approaches, which are presented as being better, and placing further natural resources on the market. Regarding this phenomenon, Humphreys (2006) points to the interest of both developing and developed countries to increase access to forests for forest industries. In the end, market-based and Western approaches erode local and indigenous communities (Aurenhammer 2013).

In contrast to this dominant approach, the author has coordinated and monitored forest projects in South America and West Africa based exclusively upon local expertise and the knowledge of subsistence peasants and indigenous peoples. These practical experiences confirm anthropologists' views (Miller 2009) of indigenous and women groups redefining prospects to improve their livelihoods and to protect themselves against developmental aggression. Local people easily contribute to cooperation projects with their own perspectives based on their knowledge, history, and context. Therefore, international cooperation could also play a useful role in directly supporting groups excluded from power to participate in the decision-making process (Acemoglu and Robinson 2012). This approach would encourage progress in democratization and pluralization (Grainger 2004) to benefit the most effective source of pressure—domestic protectionist groups (Grainger and Konteh 2007).

Thus, a key element needed to support local and indigenous communities is participation (Parrotta and Trosper 2012). Encouraging their involvement through participatory workshops allows them to design cooperation projects based upon their self-reliance and the capacity to develop their own innovations and solutions (Aurenhammer 2013; Parrotta and Trosper 2012). Concerning this important issue, the guidance and safeguards for REDD+ provide an appropriate frame to ensure the participation of local peoples in the implementation of policies to avoid deforestation. Thus, measures seeking to inhibit deforestation should promote and support "respect for the knowledge and rights of indigenous peoples and members of local communities," as well as "the full and effective participation of relevant stakeholders, in particular indigenous peoples and local communities" (decision 1/CP.16 of the UNFCCC, 2010). Indeed, the REDD+ initiative has mobilized financial resources amounting to USD 1.1–2.7 billion annually (Angelsen et al. 2018c). This amount is perceived as too small from the perspective of results-based payments. However, this funding, if well directed, would help strengthen thousands of local communities around the Global South. In this respect, Fletcher et al. (2016) proposed to take advantage of REDD+ funding to support the access and control of communities to resources that they can then manage as commons, thereby promoting community interests against extractivism.

The alternative path to forest-related cooperation discussed here enables socio-economic approaches well beyond classical Western forestry techniques. For instance, supporting rural women would definitely enhance the efficiency of policies to avoid deforestation (Fig. 12.3). It was already mentioned above the close relationship of women to natural resources. Women collect diverse forest products and simultaneously carry out accurate monitoring, which, in turn, results in a good forest condition (Agarwal 2000). Nonetheless, due to the patriarchal and unequal organization of peasant societies, women usually have less access than men to private property and depend on communal resources for subsistence (Martínez Alier 2011). Moreover, they are also excluded from decision-making processes (Agarwal 2000). Rupa Ghale, a member of the Progressive Women's Group in Nepal, summarized this issue: "because of our patriarchal culture, women were and still are seen to occupy a lower position than the men. Society does not expect a woman to put her views in front of men [...] But today, women are not only working for the improvement of our role and status in society. We are also actively struggling for our right to our land" (in Sherpa 2011). This is an unjust context that also prevents incorporating women's knowledge of biodiversity and traditional silviculture into practical measures that support sustainability (Agarwal 2000). NGOs and donors can play an important role in supporting women's forest-related activism in local organizations (Agarwal 2000), thereby ensuring and encouraging their participation

Fig. 12.3 Women and children attending a participatory workshop in Sogliboi (Kintampo, Ghana). Supporting women should constitute an important component of policies to avoid deforestation. (Photo: Jesús García Latorre)

in workshops and pressuring governments to make rural societies more inclusive toward women. The following activities of relevance for women can also be reinforced in the frame of forest projects: the establishment of vegetable gardens, handicraft activities, the commercialization of products at local markets, measures against domestic violence, etc. These and other approaches, consequently and constantly applied, should aim at improving the situation of rural women. As Martínez Alier (2011) emphasizes, "what is needed is new communitarian institutions based on eco-feminist economics and values, rather than a return to traditions of discrimination against women"; he adds that "One cannot but agree."

12.7 Prospects for Nonviolent Political Struggles Toward Sustainability

The examples described in Sect. 12.5 above are only some of the many cases portraying the dramatic situation of rural communities in developing countries. The assault on their resources is further illustrated by the issue of land grabbing. More than 220 million hectares of land have been bought or leased since the beginning of the century by financial institutions (Bommert 2012). Land grabbing features a global dimension (Fontana 2019) and has a troubling forecast. "Over the next few decades," emphasizes Fred Pearce (2012), "I believe land grabbing will matter more, to more of the planet's people, even than climate change. The new land rush looks increasingly like a final enclosure of the planet's wild places." More than half of this land appropriation has been carried out by the biofuel industry (Bommert 2012). It is expected that new alliances between multinational corporations, governments, and producers in the Global South, aiming at increasing control over the production and distribution of biofuels, will accelerate deforestation and further marginalize rural communities (Dauvergne and Neville 2010).

Indeed, the International Finance Corporation (IFC) of the World Bank Group funds, through financial intermediaries, business initiatives that promote land grabbing (Inclusive Development International 2016a). For instance, in Gabon, the Singaporean company Olam received a USD 228 million loan from one of these intermediaries to develop oil palm plantations. This firm has already cleared at least 19,000 hectares of mature forests within a concession provided by the Gabonese government (Inclusive Development International 2017). Furthermore, in Bangladesh, the National Thermal Power Corporation (NTPC) has proposed a coal plant that threatens to devastate the Sundarbans, an area that encompasses the world's largest remaining mangrove forest and a population of around two million people that depend on this unique ecosystem. Despite the grim track record of the NTPC, including projects dislocating indigenous peoples, polluting water, and deforesting, the company has benefitted from funding provided by the IFC through infrastructure and commercial banks (Inclusive Development International 2016b).

Not surprisingly, the acknowledged historian Josep Fontana (1931–2018) regarded the usurpation of land belonging to local communities as the most spectacular aspect of recent predatory capitalism and compared it convincingly with the expropriation of common lands that European peasant communities underwent in the nineteenth century (Fontana 2019). Pearce (2012), in the same vein, refers to this phenomenon as "a last roundup on the global commons." Indeed, peasants' protests against the development programs of governments and international corporations (Fontana 2013, 2019) represent a struggle against an inherent attribute of capitalism: accumulation by dispossession (Harvey 2005; Lessenich 2017; Brand and Wissen 2017).

This context might constitute an insurmountable challenge for REDD+ as the frame for policies to avoid deforestation. Notwithstanding the achieved successes of the initiative, like the more than 350 projects covering 43 million of hectares and the mobilization of considerable financial resources (Angelsen et al. 2018c), evidence suggests that REDD+ might be reaching limits that will be difficult to overcome. As commented above, development programs for large-scale conversion overlap with REDD+ implementation (Thu Thuy et al. 2018). While international financial resources for such programs are available, the funding for REDD+ falls short (Angelsen et al. 2018c). Similarly, subsidies in Brazil and Indonesia for agribusiness and industrial logging amount overall to USD 40 billion annually (Angelsen et al. 2018c), making opportunity costs very difficult to counteract.

Moreover, there is a subtle aspect limiting the effectivity of REDD+, namely, the development of the initiative within the dominant paradigm of ecological modernization. This approach assumes the feasibility of the ecological modernization of capitalism. However, it prevents the analysis of issues at the heart of the environmental crisis like capital accumulation by dispossession, territorial expansion, profit-orientation, externalization, and social dominance relationships (Harvey 2005, Lessenich 2017, Brand and Wissen 2017). Consequently, despite the multiple commitments to sustainability, extractivist politics dominate (Brand and Wissen 2017), as the following example illustrates. Bolivia, supported by Brazil, Peru, Paraguay, and Uruguay, is pushing for the construction of a 3700 km railway across South America, the Central Bioceanic Corridor. This project focuses particularly on cargo transport and seeks to encourage trade with Asia. It is budgeted at no less than USD 10,000 million and has caught the interest of European and Russian firms (Magariño 2018). Clearly, the presence of a new railway will potentially stimulate agribusiness and extractive industries, thus further exacerbating deforestation.

Against this background, it seems indispensable to challenge political and economic elites and to fight for natural resources and social equality to seek an ecosocial transition (Brand and Wissen 2017). These disputes trigger conflicts like the ones involving a myriad of local communities in the Global South (Martínez Alier 2011; Brand and Wissen 2017). Their protests bring to the forefront the political dimension of rural issues like deforestation and land inequality (Fontana 2013). In fact, politics constitutes the most essential force backing capital expansion and accumulation (Harvey 2005; Fontana 2013; Lessenich 2017), which are economic aspects at the core of the deforestation problem. Consequently, the political struggle of grassroots movements is unavoidable (Fontana 2013; Brand and Wissen 2017).

The kind of struggle addressed here is, obviously, a nonviolent one. Violence leads only to pain and suffering and definitely deserves to be rejected. Moreover, nonviolent contestation attracts more supporters and debilitates the adversary more efficiently. Nonviolence has, therefore, a better chance to succeed (Chenoweth and Stephan 2011).

Local-level protests by the victims of "development" are certainly very important. Nonetheless, as emphasized by Aurenhammer (2013), "to reform policies and institutions that are driving deforestation and social exclusion […] land reforms are also vital." Hereafter, use will be made of the experience gained by agrarian movements for land reform (Petras and Veltmeyer 2007). Indeed, both deforestation and land inequality are problems that affect rural people and share a common background. For instance, Pimple et al. (2005) point to "the legal and institutional mechanisms that are designed to deprive the majority of land," a view that is also relevant to understanding deforestation. Thus, developing alternative land policies implies challenging the established structure (Pimple and Sethi 2005). This is why the support of organized movements and broad coalitions contesting the existing regime (Acemoglu and Robinson 2012) is essential to push for reforms that benefit local and indigenous communities (Pimple and Sethi 2005). Moreover, Harvey (2005) indicates that in the struggles against accumulation by dispossession the majority of the population has to respond in class terms to the "ever-increasing upper-class power." Therefore, Petras et al. (2007) emphasize the relevance of class organization for agrarian movements. The only way for them to bring about fundamental changes is to develop their own political organizations and build alliances with other groups to influence, pressure, or overthrow the state (Petras and Veltmeyer 2007).

Numerous poor local and indigenous communities around the Global South sustainably manage their natural resources and oppose extractivism. They confirm that inspiration and hope frequently come from those who, though being poor, have cultural wealth beyond all measure (Miller 2009). Their initiatives seek equity, welfare, respect for life, and solidarity and achieve bigger impacts than projects conceived by official institutions (Smith and Max-Neef 2011). As long as indigenous and local communities do not have a strong stake in the implementation of policies to avoid deforestation, the deforestation problem will remain elusive. As emphasized by Martínez Alier (2011), the grassroots movements of *the environmentalism of the poor* are a reason for optimism. These movements represent the main social force that is searching for allies to move the economy in a more fair and sustainable direction. Certainly, alternatives are born from resistance (Martínez Alier 2011). Let us support them!

References

abc color (2018) Gobierno deroga decreto por el cual Cartes arrasó bosques. https://www.abc.com.py/edicion-impresa/economia/gobierno-deroga-decreto-por-el-cual-cartes-arraso-bosques-1733002.html. Accessed 1 Oct 2019

Acemoglu D, Robinson JA (2012) Why nations fail. The origins of power, prosperity, and poverty. Profile Books, London

Affif S (2009) An Indonesian overview. Indigenous peoples' writing on forest management: a counter discourse? In: Kleden O, Indradi Y, Chidley L (eds) Forest for the future. Indigenous forest management in a changing world. AMAN & DTE, Jakarta, pp 263–274

Agarwal B (2000) Conceptualising environmental collective action: why gender matters. Camb J Econ 24:283–310

Amazon Watch (2018) Complicity in destruction: how northern consumers and financiers sustain the assault on the Brazilian Amazon and its peoples. Amazon Watch, Washington

Angelsen A et al (2018a) Introduction. REDD+ enters its second decade. In: Angelsen A et al (eds) Transforming REDD+. Lessons and new directions. CIFOR, Bogor Barat, pp 1–13

Angelsen A et al (eds) (2018b) Transforming REDD+. Lessons and new directions. CIFOR, Bogor Barat

Angelsen A et al (2018c) Conclusions. Lessons for the path to a transformational REDD+. In: Angelsen A et al (eds) Transforming REDD+. Lessons and new directions. CIFOR, Bogor Barat, pp 201–214

Aranha C (2019) Governo foi alertado pelo Ministério Público três dias antes de "dia do fogo". https://revistagloborural.globo.com/Noticias/Politica/noticia/2019/08/governo-foi-alertado-pelo-ministerio-publico-tres-dias-antes-de-dia-do-fogo.html. Accessed 1 Oct 2019

Aurenhammer PK (2013) Development cooperation policy in forestry from an analytical perspective. Springer, Dordrecht

Bayona E (2018) Asesinados 437 ambientalistas y activistas por Derechos Humanos en dos años. Público, https://www.publico.es/sociedad/asesinados-437-ambientalistas-activistas-pro-derechos-humanos-anos.html. Accessed 1 Oct 2019

Benjaminsen T (1997) Is there a fuelwood crisis in rural Mali? GeoJournal 43:163–174
Berkes F, Folke C (1998) Linking social and ecological systems for resilience and sustainability. In: Berkes FC (ed) Linking social and ecological systems: management practices and social mechanisms. Cambridge University Press, Cambridge, pp 1–25
Bommert W (2012) Bodenrausch. Die globale Jagd nach den Äckern der Welt. Eichborn, Köln
Brand U, Wissen M (2017) Imperiale Lebensweise. Zur Ausbeutung von Mensch und Natur im globalen Kapitalismus. Oekom, München
Chenoweth E, Stephan MJ (2011) Why civil resistance works. The strategic logic of nonviolent conflict. Columbia University Press, New York
Crewe E, Harrison E (1998) Whose development? An ethnography of aid. Zed Books, London
Cronkleton P, Barry D, Pulhin JM, Saigal S (2010) The devolution of management rights and the co-management of community forests. In: Larson AM et al (eds) Forests for people. Community rights and forest tenure reform. Earthscan, London, pp 43–68
Cuentas S (2015) Soy pobre y analfabeta, pero lucharé por nuestras montañas El País, https://elpais.com/elpais/2015/03/17/planeta_futuro/1426588446_691506.html. Accessed 1 Oct 2019
Dauvergne P, Neville KJ (2010) Forests, food, and fuel in the tropics: the uneven and ecological consequences of the emerging political economy of biofuels. J Peasant Stud 37(4):631–660
Davey I (2009) Environmentalism of the poor and sustainable development: an appraisal. JOAAG 4(1):1–10
Dooley K, Gupta A (2017) Governing by expertise: the contested politics of (accounting) for land-based mitigation in a new climate agreement. Int Environ Agreements 17(4):483–500
Easterly W (2006) The white man's burden. Why the west's efforts to aid the rest have done so much ill and so little good. The Penguin Press, New York
Fay C (2009) An international overview. Indigenous natural resource management systems at the crossroads. In: Kleden O, Indradi Y, Chidley L (eds) Forest for the future. Indigenous forest management in a changing world. AMAN & DTE, Jakarta, pp 277–294
Fenly N (2011) Forests and indigenous women in Tuapi: "return to Auhbi Piakan". In: Alangui V, Subido G, Tinda-an R (eds) Indigenous women, climate change and forests. Tebtebba, Baguio City, pp 41–74
Fletcher R, Dressler W, Büscher B, Anderson Z (2016) Questioning REDD+ and the future of market-based conservation. Conserv Biol 30(3):673–675
Fontana J (2013) El futuro es un país extraño. Una reflexión sobre la crisis social de comienzos del siglo XXI. Pasado y Presente, Barcelona
Fontana J (2019) Capitalismo y democracia 1756–1848. Cómo empezó este engaño. Crítica, Barcelona
Forest Peoples Programme (2012) Forest peoples. Numbers across the world. Forest Peoples Programme, Moreton-in-Marsh
Gabay M, Alam M (2017) Community forestry and its mitigation potential in the Anthroponcene: the importance of land tenure governance and the threat of privatization. Forest Policy Econ 79:26–35
García Latorre J, García Latorre J (2012) Globalization, local communities, and traditional forest-related knowledge. In Parrotta J, Trosper R L (eds) Traditional forest-related knowledge. Sustaining communities, ecosystems and biocultural diversity. Springer, Dordrecht, pp 449–490
García Latorre J, García Latorre J, Sánchez Picón A (2001) Dealing with aridity: socio-economic structures and environmental changes in an arid Mediterranean region. Land Use Policy 18:53–64
Global Witness (2009) Trick or treat? REDD, development and sustainable forest management. Global Witness, London
Global Witness (2017a) Defenders of the earth. Global killings of land and environmental defenders in 2016. Global Witness, London
Global Witness (2017b) Honduras. El lugar más peligroso para defender el planeta. Global Witness, London

Global Witness (2018) At what cost? Irresponsible business and the murder of land and environmental defenders in 2017. Global Witness, London
Godelier M (1981) Instituciones económicas. Editorial Anagrama, Barcelona
Grainger A (2004) Societal change and the control of deforestation in Thailand. Int J Sustain Dev World Ecol 11:364–379
Grainger A, Konteh W (2007) Autonomy, ambiguity and symbolism in African politics: the development of forest policy in Sierra Leone. Land Use Policy 24:42–61
Grainger A, Malayang BS (2006) A model of policy changes to secure sustainable forest management and control of deforestation in the Philippines. Forest Policy Econ 8:67–80
Greenpeace (2018) Desmontes S.A. Parte 4. La responsabilidad empresaria y gubernamental en la violación de la Ley de Bsques en Chacho. Greenpeace, Buenos Aires
Griffiths T (2007) Seeing "RED"? "Avoided deforestation" and the rights of indigenous peoples and local communities. Forest People Programme, Moeton-in-Marsh
Guereña A, Rojas Villagra L (2016) Yvy Jára. Los dueños de la tierra en Paraguay. OXFAM, Asunción
Guiang ES, Borlagdan SB, Pulhin JM (2001) Community-based forest management in the Philippines: a preliminary assessment. Institute of Philippine culture, Ateneo de Manila University, Quezon City
Hansen MC et al (2013) High-resolution global maps of 21st-century forest cover change. Science 342:850–853
Hargita Y, Günter S, Köthke M (2016) Brazil submitted the first REDD+ reference level to the UNFCCC – implications regarding climate effectiveness and cost-efficiency. Land Use Policy 55:340–347
Harvey D (2005) A brief history of neoliberalism. Oxford University Press, Oxford
Higgonet E, Hurowith G, Bellantonio M, Lapidus D (2016) Palm oil's black box. How Singapore's national wealth fund has fuelled a secret vehicle for deforestation from Gabon to Indonesia. Mighty, Washington
Holling CS, Berkes F, Folke C (1998) Science, sustainability and resource management. In: Berkes F, Folke C (eds) Linking social and ecological systems: management practices and social mechanisms. Cambridge University Press, Cambridge, pp 342–362
Humphreys D (2006) Public goods, neoliberalism and the crisis of deforestation. Paper presented to the annual conference of the British International Studies Association, University College Cork, Cork, 18–20 Dec 2006
Inclusive Development International (2016a) Outsourcing development: lifting the veil on the World Bank Group's lending through financial intermediaries. IDI, Asheville
Inclusive Development International (2016b) Bankrolling India's dirty dozen. IDI, Asheville
Inclusive Development International (2017) Unjust enrichment: how the IFC profits from land grabbing. IDI, Asheville
Jashimuddin M, Inoue M (2012) Community forestry for sustainable forest management: experiences from Bangladesh and policy recommendations. FORMATH 11:133–166
Kariuki PM, Njoka JT, Saitabau CL, Saitabu HS (2016) Forest governance, livelihoods and resilience: the case of Loita forest (Entime e Naimina enkiyio), Narok County, Kenya. In: van Dijk H (ed) Bose P. Springer, Dryland forests, pp 117–138
Keenan RJ et al (2015) Dynamics of global forest area: results from the FAO Global Forest Resources Assessment 2015. For Ecol Manag 352:9–20
Kimaren S, Riamit O (2011) Indigenous women in forest management: the Maasai women of Naimina Enkiyio forest, Southern Kenya. In: Alangui WV, Subido G, Tinda-an R (eds) Indigenous women, climate change and forests. Tebtebba Foundation, Baguio City, pp 165–206
Kretschmer R (2014) Das Massaker von Marina Kue (Curuguaty). Landkonflikte und Verfolgung von BäuerInnen in Paraguay. FIAN Fact Sheet 2014/5, Köln
Lambin EF et al (2001) The causes of land-use and land-cover change: moving beyond the myths. Glob Environ Chang 11:261–269

Larson AM, Barry D, Ram Dahal G (2010) Tenure change in the Global South. In: Larson AM, Barry D, Ram Dahal G, Pierce Colfer CJ (eds) Forests for people: community rights and forest tenure reform. Earthscan, London, pp 3–18
Lasco RD, Pulhin JM (2006) Environmental impacts of community-based forest management in the Philippines. Int J Environ Sustainable Dev 5(1):46–56
Lessenich S (2017) Neben uns die Sintflut. Die Externalisierungsgesellschaft und ihr Preis. Hanser, Berlin
Magariño JF (2018) Constructoras españolas entran en la carrera por construir el tren que cruzará Sudamérica. El País, https://cincodias.elpais.com/cincodias/2018/09/10/companias/1536606993_991235.html. Accessed 1 Oct 2019
Martínez Alier J (1994) De la economía ecológica al ecologismo popular. Icaria, Barcelona
Martínez Alier J (2011) El ecologismo de los pobres. Conflictos ambientales y lenguajes de valoración. Icaria, Barcelona
McAfee K (2012) The contradictory logic of global ecosystem services markets. Dev Chang 43(1):105–131
Meiners U, Rösener W (eds) (2004) Allmenden und Marken vom Mittelalter bis zur Neuzeit. Museumsdorf Cloppenburg, Cloppenburg
Miller B (2009) Cultural anthropology, 5th edn. Prentice Hall
Montagnini F, Jordan CF (2005) Tropical forest ecology. The basis for conservation and management. Springer, Berlin
Murali KS (2002) Joint forest management in India and its ecological impacts. Environ Manag Health 13(5):512–528
Nepstad D et al (2006) Inhibition of Amazon deforestation and fire by parks and indigenous lands. Conserv Biol 20(1):65–73
Nepstad D et al (2014) Slowing Amazon deforestation through public policy and interventions in beef and soy supply chains. Science 344:1118–1123
Ola Kleden E, Indradi Y, Chidley L (eds) (2009) Forest for the future. Indigenous forest management in a changing world. AMAN & DTE, Jakarta
Parrotta J, Trosper RL (eds) (2012) Traditional forest-related knowledge. Sustaining communities, ecosystems and biocultural diversity. Springer, Dordrecht
Paudel NS, Monterroso I, Cronkleton P (2010) Community networks, collective action and forest management benefits. In: Larson AM, Barry D, Ram Dahal G, Pierce Colfer CJ (eds) Forests for people: community rights and forest tenure reform. Earthscan, London, pp 116–136
Pearce F (2012) The land grabbers. The new fight over who owns the earth. Beacon Press, Boston
Persson R (2000) Assistance to forestry: what have we learnt? Int For Rev 2(3):218–224
Petras J, Veltmeyer H (2007) The 'development state' in Latin America: whose development, whose state? J Peasant Stud 34(3–4):371–407
Pimple M, Sethi M (2005) Occupation of land in India: experiences and challenges. In: Moyo S, Yeros P (eds) Reclaiming the land. Asia and Latin America, Zed Books, The resurgence of rural movements in Africa, pp 235–256
Porter-Bolland L et al (2012) Community managed forests and forest protected areas: an assessment of their conservation effectiveness across the tropics. For Ecol Manag 268:6–17
Pretty J (2003) Social capital and the collective management of resources. Science 302:1912–1914
Radkau J (2012) Umwelt und Macht. Eine Weltgeschichte der Umwelt. C. H. Beck, München
Ravindranath NH, Murali KS, Sudha S (2006) Community forestry initiatives in Southeast Asia: a review of ecological impacts. Int J Environ Sustainable Dev 5(1):1–11
Regadas L (2017) Quieren dejar claro que en Honduras pueden matar a quien les dé la gana. Público, https://www.publico.es/internacional/berta-caceres-quieren-dejar-claro.html. Accessed 1 Oct 2019
Riquelme Q (2005) Los sin tierra en Paraguay. Conflictos agraios y movimiento campesino. CLACSO, Buenos Aires

Rukunuddin M (2003) International development assistance in the forestry sector: the rhetoric versus reality. World Forestry congress XII, http://www.fao.org/3/XII/0350-C5.htm. Accessed 1 Oct 2019

Scott JC (1998) Seeing like a State. Yale University Press, New Haven

Sherpa T (2011) Indigenous women in Khasur and Kalleri villages of Nepal: traditional knowledge and adaptation strategies in the face of climate change. In: Alangui WV, Subido G, Tinda-an R (eds) Indigenous women, climate change and forests. Tebtebba Foundation, Baguio City, pp 93–126

Smith Ph B, Max-Neef M (2011) Economics unmasked. From power and greed to compassion and the common good. Green Books, Foxhole

Song X-P et al (2018) Global land change from 1982 to 2016. Nature 560:639–643

Syahruji A, Kiyu B (2009) The Kiyu Dayak indigenous community Meratus, South Kalimantan. In: Kleden O, Indradi Y, Chidley L (eds) Forest for the future. Indigenous forest management in a changing world. AMAN & DTE, Jakarta, pp 101–141

Thu Thuy P et al (2018) Strategic alignment. Integrating REDD+ in NDCs and national climate policies. In: Agelsen A et al (eds) Transforming REDD+. Lessons and new directions. CIFOR, Bogor Barat, pp 69–80

Trakansuphakorn P, Kampholkul T (2010) Knowledge and practice on rotational farming of Pgaz K'Nyau (Karen) people, Hin Lad Nai Community in Northern Thailand. In: Tauli-Corpuz V, Enkiwe-Abayao L, de Chavez R (eds) Towards an alternative development paradigm: indigenous peoples' self-determined development. Tebtebba Foundation, Baguio City, pp 249–329

Vedeld P, Angelsen A, Bojö J, Sjaastad E, Berg GK (2007) Forest environmental incomes and the rural poor. Forest Policy Econ 9:869–879

Weisse M, Ruiz S (2019) What can Global Forest Watch tell us about the fires in Brazil? https://blog.globalforestwatch.org/fires/what-can-global-forest-watch-tell-us-about-the-fires-in-brazil. Accessed 1 Oct 2019

Williams M (2003) Deforesting the earth. The University of Chicago Press, Chicago

World Commission on Environment and Development (1987) Our common future. Oxford University Press, Oxford

Worster D (1988) Doing environmental history. In: Worster D (ed) The ends of the earth. Perspectives on modern environmental history. Cambridge University Press, Cambridge, pp 289–307

Part III
Participatory Biodiversity Conservation: The Challenges Ahead

Chapter 13
Prospects for Participatory Biodiversity Conservation in the Contemporary Crisis of Democracy

Cristina Baldauf

13.1 New Routes and Frameworks Are Necessary to Address Both Conservation Goals and Human Well-Being

Despite the lack of consensus on whether new routes and approaches for conservation are necessary to halt biodiversity loss (Godet and Devictor 2018), current evaluations and future scenarios show humanity will likely fail to meet the 2020 Aichi Biodiversity Targets (IPBES 2019). Moreover, although conservation science and practice have made great progress in steering society toward ecological sustainability, the importance of social justice has not been recognized as an equally important goal (Salomon et al. 2018). For instance, the establishment of protected areas has caused innumerous violations of human rights worldwide (Tauli-Corpuz et al. 2018), displacing local populations and/or creating new regulations with negative impacts on the food security and livelihoods of local communities (Chaps. 4 and 5, this volume). Within this context, the increase in number and/or area of protected areas to achieve CBD's Aichi Target 11 would represent significant threats to food security and local livelihoods in Africa, Asia, and Latin America due to reductions in the access to wild foods and potential restrictions on traditional land-use systems (Brondizio et al. 2016) (Fig 13.1).

A critical analysis of global conservation projects at the scale of the Half-Earth project (see Chaps. 2 and 3) is of the utmost importance since its influence on global environmental governance is growing (Ellis 2019; Ellis and Mehrabi 2019). A recent study concluded that if Half-Earth were carried out, it would have serious consequences for food security, namely, a loss of 15–31% of cropland, 10–45% of pastureland, 23–25% of non-food calories, and 3–29% of food calories (Mehrabi et al. 2018). The same study also revealed that countries with the highest numbers

C. Baldauf (✉)
Departamento de Biociências, Universidade Federal Rural do Semi-Árido, Mossoró, Rio Grande do Norte, Brazil
e-mail: crisbaldauf@ufersa.edu.br

C. Baldauf (ed.), *Participatory Biodiversity Conservation*,
https://doi.org/10.1007/978-3-030-41686-7_13

Fig. 13.1 Landscape mosaic resulting from swidden-fallow agriculture in the Brazilian Atlantic forest. Photo: Cristina Baldauf

of undernourished people would suffer substantial losses of food calories; on the other hand, a land-sharing approach would dramatically minimize the trade-offs between conservation and food production (Mehrabi et al. 2018). Consequently, a reorientation of conservation efforts, away from an exclusive focus on protected areas toward a landscape approach focusing on the promotion of biodiversity-friendly agricultural methods, is imperative (Perfecto et al. 2009).

The adoption of agroforestry systems is increasingly considered as an effective tool for biodiversity conservation since these systems can sustain high levels of species richness and provide several ecosystem services, as well as increase connectivity in fragmented landscapes (Perrings et al. 2006; Jose 2012; Sistla et al. 2016; Haggar et al. 2019). Furthermore, productive restoration using agroforestry and agroecological models, as presented in Chap. 8, can contribute to the restoration of the structure and functions of the original ecosystem while generating economic benefits and promoting food security/sovereignty (Fig 13.2). This productive restoration framework is also in line with the forest landscape restoration approach (FLR) of the Bonn Challenge,[1] which aims to both restore forest ecological integrity and improve human well-being through multifunctional landscapes. Similarly, locally driven approaches to fisheries management, such as the experience of mac-

[1] The Bonn Challenge is a global effort to restore 150 million hectares of the world's degraded and deforested lands by 2020 and 350 million hectares by 2030. It is overseen by the Global Partnership on Forest Landscape Restoration, with the International Union for Conservation of Nature as its Secretariat. See http://www.bonnchallenge.org

Fig. 13.2 Agroforestry systems (AS) in different countries. (**a**) AS developed by a *Quilombola* small farmer in a sustainable development reserve in southeastern Brazil. (**b**) AS in Belém, Brazilian Amazon. (**c**) Detail of the fruiting of the cocoa tree in an AS. (**d**) AS on Sumatra Island, Indonesia. (**e**) A homegarden in Canary Islands, Spain. Photos: Felipe Augusto Zanusso Souza (**a**), Cristina Baldauf (**b**, **e**), Deyvid Oliveira (**c**), Intan Fardinatri (**d**)

roalgae mariculture presented in Chap. 6, constitute an alternative to the creation of artificial conservation islands through strictly protected areas in aquatic environments (Stephenson et al. 2014). Such local initiatives can be expanded with the goal of achieving socioecological benefits at a regional scale (Chap. 7), especially if certain institutional and design principles, such as collective-choice arrangements and nested enterprises, are taken into account (Ostrom 2009; Cox et al. 2010).

Notwithstanding the polarization that has dominated the debate in biodiversity conservation (Chap. 2), most of the community believes that it is possible to reconcile the human welfare and cultural heritage of Traditional Peoples and Local Communities (TPLC) with conservation goals. To this end, a biocultural approach is considered a useful model for biodiversity conservation and monitoring since it represents a synthesis of theories from multiple fields, drawing on pluralistic values and addressing the erosion of both cultural and biological diversity (Gavin et al. 2015, 2018).

The principles of biocultural approaches to conservation are as follows: *1. acknowledge that conservation can have multiple objectives and stakeholders; 2. recognize the importance of intergenerational planning and institutions for long-term adaptive governance; 3. recognize that culture is dynamic, and this dynamism shapes resource use and conservation; 4. tailor interventions to the social-ecological context; 5. devise and draw upon novel, diverse, and nested institutional frameworks; 6. prioritize the importance of partnership and relation building for conservation outcomes; 7. incorporate the distinct rights and responsibilities of all parties;* and *8. respect and incorporate different worldviews and knowledge systems into conservation planning* (Gavin et al. 2015). Most of these principles are directly or indirectly related to the participation of Traditional Peoples and Local Communities (TPLC) in conservation initiatives. However, the term *participation* is interpreted in many different ways, and only certain types of participation can lead to the full involvement of TPLC. Some of the challenges and opportunities for equity and more inclusive participation by nonprivileged groups, as well as to the adequate representation of their views, were discussed in several chapters of this book, and I will try to build on their findings in the following two sections.

13.2 Future Prospects for Participatory Approaches

Biodiversity conservation has historically favored a top-down approach characterized by the establishment of protected areas and an imposition of natural resource management rules by formal institutions (Matarrita-Cascante et al. 2019). Although this trend is still dominant in the field, approaches that promote joint conservation and development strategies between land management authorities and local peoples have substantially increased and gained legitimacy since the 1980s (Bixler et al. 2015, Chap. 2). A myriad of approaches has since been employed, which were covered in the precedent chapters under the conceptual umbrella of *participatory biodiversity conservation*, since these bottom-up approaches support, at least in theory, the principles of participatory democracy. Furthermore, *communities* do not represent isolated agents, as they are embedded in social multiscale systems (Berkes 2004). Thus, the term *participatory* rather than *community-based* is broader and encompasses the experiences presented in this book.

Although participatory processes have gained prominence over the past few decades, they are also subject to criticism and controversy, in part because the goals of community development and conservation are not necessarily compatible (Berkes 2004; López-Bao et al. 2017, Chap. 9, Chap. 12). As a result, and also due to the variety of bottom-up approaches, some experiences are positive for biodiversity conservation, while others are not. Thus, it makes little sense to ask questions such as "does participatory biodiversity conservation work, or not?", and it becomes more interesting to ask under what conditions it may or may not work (Berkes 2009).

A number of design principles and characteristics of successful community-based approaches for managing common pool resources (CPRs) have been identified, although many are context-specific (Ostrom 2009; Gruber 2010; Cox et al. 2010; Brooks et al. 2013). Perhaps the most widely used research framework for understanding the limits and possibilities of conservation and the management of CPRs is Elinor Ostrom's design principles for robust performance (Ostrom 1990). This framework has been modified or expanded over the years through critical analysis and empirical studies as well as integrated with other theories to develop suitable tools for designing sustainable CPRs institutions (Wittayapak and Dearden 1999; Quinn et al. 2007; Ostrom 2009; Gruber 2010; Araral 2014; Cox et al. 2016; Bastakoti and Shivakoti 2017; Singleton 2017).

With specific regard to community participation in biodiversity conservation—the central theme of this book—a key question is: *which types of participation result in mutual benefits for biodiversity and local livelihoods*? There is no obvious answer to this question, since participation is a complex term and it is interpreted in different ways depending on socioeconomic, cultural, political, and ecological context (Bixler et al. 2015). Still, it is possible to identify some preconditions and principles related to community participation that have often been associated with successful initiatives. Gruber (2010) grouped the results of publications from 23 research groups based on broad organizational principles and associated key characteristics related to effective community participation and mobilization. This author points out that public participation should empower citizens, raise knowledge levels, and involve a great variety of stakeholders. To this end, he considers it extremely important to enable stakeholder participation at all stages of the initiative and provide opportunities for sharing knowledge and collaborative learning about social-ecological systems (Gruber 2010).

Brooks et al. (2013) also evaluated the factors associated with the success or failure of community-based conservation projects and concluded that project design, particularly capacity building in local communities, is critical to success. Another synthesis on the factors that should be considered for the success of participatory conservation initiatives was recently published by Matarrita-Castante et al. (2019). Among the factors listed by the authors, based on a broad literature review, are (a) principles of common property rights; (b) social and economic benefits derived from community participation in the management of natural resources and proactive conservation behaviors; (c) understanding the ecology and biology of the resource in question; (d) defined physical/geographical boundaries for the resource management; (e) understanding the relationship between the community and the managed resource(s), particularly with regard to the economic and cultural role of a resource in the local livelihoods; and (f) power transfer to communities through a legal framework.

Although the factors listed above do not represent a list of requirements for success, the case studies presented in this book suggest they are important points to consider improving conservation effectiveness and local livelihoods. On the other hand, both Gruber (2010) and Matarrita-Castante (2019) do not assign much importance to the political dimension of socioecological systems and their influence on the success (or lack thereof) of participatory conservation, whereas Brooke et al. (2013) go further and claim that national context does not influence project outcomes. However, a contrasting perspective will be presented in the last topic of this chapter.

13.3 Beyond Participation Typologies: The Search for Appropriate Participation Levels and Governance Models

> There is a critical difference between going through the empty ritual of participation and having the real power needed to affect the outcome of the process. Sherry Arnstein (1969)

> The governmental siege is behind us, it didn't work and never will. We're following paths and routes that neither exist on maps nor in the images of satellites and are only found in the thoughts of our elders. EZLN - Zapatista Army of National Liberation (2019)

The intense theoretical and methodological debate about participation in Political Sociology and later in Political Ecology has resulted in the proposition of several typologies that aim to reveal a gradient of types of participation, such as those of Arnstein (1969), Pretty (1995), and White (1996). Among these, Arnstein (1969) is possibly the best known. In this typology, eight rungs of a ladder represent the citizen's power to determine the end product by participating in a government plan or program. At the bottom of the ladder is the *manipulation* step and at the top the *citizen control* step. The eight steps are divided into three broad categories: nonparticipation, tokenism, and citizen power. Pretty (1995) proposed a typology of participation focused on citizen control over the state and motivations for participation. At the lowest level is manipulative participation, while at the top is self-mobilization. Finally, White's (1996) typology is based on four levels of participation: nominal, instrumental, representative, and transformative. Although not proposed in the ladder model, it is possible to recognize a participation gradient where nominal participation seeks to legitimize state action, while transformative participation is grounded in community empowerment in decision-making. The typologies presented are a simplification, as many forms of participation can occur simultaneously and over time in a given project or process; however, they are useful to illustrate the gradations of citizen participation and power distribution.

It is not difficult to identify similarities in the three typologies presented. Expressed conditions of no or little participation are at their bases, while the top rungs represent democratic forms of participation. Another similarity is that of the normative character of these typologies, as a progression toward more *genuine* forms of participation is evident. On the other hand, the endpoints of these are rather different, since citizen control goes much further than self-mobilization or what transformative participation may suggest (Cornwall 2008).

An underlying premise of participatory biodiversity conservation is that the resulting decisions will be representative, legitimate, and equitable, based on the principle of participation as a fundamental democratic right for local communities. However, the assumption that full citizen power should be the ultimate goal of projects involving social participation has recently been questioned. For Haklay et al. (2018), even though there are likely benefits in terms of participants' engagement and involvement in a project, there should not be such a strong value judgment on the position that a specific project takes. He believes that, in contrast to Arnstein's ladder, participation should be valued at many levels—from occasional contribution to deep engagement. In a similar vein, Chilvers and Kearnes (2015) believe there are multiple realities of participation, which may not be restricted to fixed and gradual participation models described in Arnstein's ladder. Instead, they understand power emerging from the relations between all people and things and working through the mediation of all collectives of participation.

Situations where full participation or full citizen power is a consequence of the state abnegating its responsibilities also can be found (Cornwall 2008). For instance, numerous IPLCs have assumed increasing responsibility in protecting forests from the advance of extractivism and the expansion of the agricultural frontier in different parts of the world (Kohler and Brondizio 2017). However, while playing the role of the state in defending biodiversity in many countries, the benefits of risking, and in many cases losing, their lives in defending their territories and associated biodiversity are limited due to the state's failure in recognizing their land rights and combating violations of their human rights (Kohler and Brondizio 2017; CIMI 2018).

There are situations where *citizen power* achieves an unforeseen step in participation typologies, as even the last step of Arnstein's ladder (citizen control) does not represent a total rupture with the state. An emblematic example of this *extra step* at the top of participation typologies is the Zapatista insurgency in Mexico (Fig 13.3). Twenty-five years ago, the indigenous peoples of Mexico in the state of Chiapas formed autonomous communities in response to attacks to their fundamental rights promoted by the accentuated neoliberal economy in the country, especially after the signing of the North American Free Trade Agreement (NAFTA). There are currently 43 territories of the Zapatista organization in the state of Chiapas, without the control of the Mexican state or political parties, which represent a means for the indigenous peoples to ensure their independence, self-organization, and own development model (Azevedo 2019). In this case, and perhaps in most cases concerning traditional populations, biodiversity conservation is not a goal in itself, but emerges from the interaction between cosmovision, ethics, knowledge, and practices. Therefore, it is necessary to avoid the temptation of framing local populations as traditional environmentalists, since it can create contradictions and frustrations for both local populations and those involved in conservation (Kohler and Brondizio 2017).

One of the greatest challenges of participatory biodiversity conservation is finding the right mix of governance systems for each situation. Perhaps one of the most successful examples, although not free from criticism, is the indigenous co-manage-

Fig. 13.3 Subcomandante Marcos speaking during "The Other Campaign" (La otra campaña), a political initiative by the Zapatista Army of National Liberation (EZLN), which intended to link the Zapatistas and other Mexican resistance groups in 2006. (Photo: Roberto Robles)

ment of protected areas in Australia (Fig. 13.4). After decades of dispossession and marginalization of indigenous peoples during the establishment of protected areas in Australia, governance and policy frameworks have been recognizing protected areas as part of indigenous *country*[2] and including indigenous people in conservation planning and management (Ross et al. 2009). Protected areas in Australia are currently run by state governments, with diverse co-management arrangements with indigenous people (see Borrini-Feyerabend et al. 2004; Ross et al. 2009; Carlsson and Berkes 2005 and Zurba et al. 2012 for co-management typologies) (Fig. 13.4).

A variety of planning and policy frameworks for co-management have been developed across Australia; however, a number of inevitable challenges and disappointments have emerged from such experiences. Most of these shortcomings are related to government-driven negotiations and outcomes (Yunupingu and Muller 2009; Rist et al. 2019) and the *stretching* (or political co-opting) of the co-management concept (Pearson and Lain 2019), but can also be derived from the power relations at the local level which can reproduce and magnify inequalities (Lane and Corbett 2005). Nevertheless, there are opportunities based on Aboriginal aspirations

[2] For Aboriginal people, *country* encompasses an interdependent relationship between an individual and their ancestral lands and seas, which is sustained by traditional ecological knowledge.

Fig. 13.4 Uluru-Kata Tjuta National Park, a UNESCO's World Heritage. The land is owned by the Anangu people and leased to Parks Australia, who manages the national park in collaboration with the traditional owners since 1985. (Photos: (**a**) Miyoshi Sakamoto, (**b**) Cristina Baldauf)

and in their holistic concept of *country* that can generate local, regional, and national benefits for indigenous and non-indigenous stakeholders (Altman 2003). Among such opportunities is the establishment of Indigenous Protected Areas (IPAs) in recent years, which have enabled indigenous people to lead planning and governance of land and sea areas and overcome some of the previous shortcomings of co-management in Australia (Rist et al. 2019).

Even though Australian co-management experiences have failed at times and sometimes the outcomes are only of symbolic importance and result in little improvement in material conditions of the marginalized indigenous populations, it is undeniable that this country has made great strides in participatory biodiversity conservation. However, such experiences cannot be transposed to other countries because of their peculiarities. Nevertheless, they serve as inspiration to all countries that have the intention to recognize indigenous rights, and perhaps we should take heed of the famous Midnight Oil song:

> The time has come to say fair's fair/To pay the rent, to pay our share/The time has come, a fact's a fact/It belongs to them, let's give it back. (5–8)

13.4 Dynamics of the Wider Political Economy Constrain Participatory Biodiversity Conservation

In the first chapter, I briefly addressed the contemporary crisis of democracy. As a result of this crisis, populations around the world have sought to better organize themselves around infinite subjects, conquering public spaces and pressuring their representatives. Their demands are becoming more complex, and the need for joint participation between representatives and those they represent is evident (Fig 13.5). Political-democratic participation, a fundamental right, is a way of legitimizing

Fig. 13.5 Demonstration by NGO groups during the climate conference in Paris, December 2015. (Photo: Jesús García-Latorre)

democracy. Citizens have the right not only to participate in decision-making that affect their collective life but also to live within authority structures that make such participation possible (Pateman 2012). Legitimacy is attributed to not only the mere respect of the rule of the majority but also the guarantee of space for participation of minorities.

Despite it now being widely accepted that citizens should be involved in environmental decision-making, especially when these decisions affect their lives, participatory biodiversity conservation is not free from criticism. A familiar critique of participatory processes is that most individuals are not sufficiently capable to participate (Pateman 2012). On the contrary, there is plenty of evidence regarding the high value of both traditional ecological knowledge (TEK) and local ecological knowledge (LEK) for biological conservation (see Reyes-García 2015 for a review). Moreover, community members are quite capable of understanding complex, and sometimes technical, issues and reaching pertinent conclusions (Pateman 2012). On the other hand, there is more solid criticism regarding the lack of legitimacy of process in environmental decision-making, such as those who take part not being truly representative of the wider community, nontransparent constitutive decisions, limited information exchange and learning, and decision-making using pseudo-democratic methods favoring powerful actors or even supporting pre-arranged outcomes (Moote et al. 1997; Apostolopoulou et al. 2012; Bixler et al. 2015; Demeritt 2015; López-Bao et al. 2017). Moreover, many governmental and international agencies incorporate participation only for instrumental reasons, or as part of legal obligations and, therefore, without changing the hierarchical dichotomy between government and people, which leads to restraints in the level of community participation (Demeritt 2015, Chap. 9).

Another recurring critique of participatory biodiversity conservation, particularly in relation to approaches based on the concept of resilience and institutional analysis of the commons, is that of its indifference regarding wider political economic dynamics, even though resilience researchers have been engaging with these concerns more recently (Brown 2014). The social subsystem of a socioecological system incorporates economics, politics, history, and ethics/worldviews (Berkes 2017). However, the extent to which a resilience perspective can address these dimensions, particularly in terms of politics, power distribution, and social justice, has been widely questioned (Leach 2008; Davidson 2010; Beymer-Farris et al. 2012; Cote and Nightingale 2012; Beymer-Farris 2013; Béné et al. 2008; Thorkildsen 2014).

Important criticisms have been made of the liberal nature of the institutional analysis of the commons. Despite the disagreements with some assumptions of the dominant neoclassical economy, Elinor Ostrom's seminal work firmly sits within the classical liberal tradition of the political economy, as it relies on the freedom of individuals to independently create the most beneficial arrangements for managing the commons (Pennington 2012). However, when considering members of IPLC as rational individuals acting on costs and benefits, Ostrom and colleagues underestimate both the social processes at play and the weight of the overall economic and political context (Zulu 2012; Singleton 2017).

For authors such as Dardot and Laval (2017), individuals are immersed in the global economy and suffer its effects or influence, raising questions about the possibilities of local institutional arrangements suffering the constraints of capitalism on their forms of organization. For them, it would be necessary to believe that "the archipelago of the commons could survive in the icy waters of the market and the state, thanks to the superior rationality of the commons in a number of very specific cases" (Dardot and Laval 2017, p. 165). These authors also criticize the lack of a generalized rationality that would allow the development of other founding institutions from the commons.

Although the criticisms presented are extremely pertinent, they seem to demand too much of an approach, even though it has laid down the foundations of the whole research agenda for the study of the commons. Elinor Ostrom never set out to propose a general principle of social organization, since throughout her research she has always defended polycentric governance based on a plurality of tenure systems and institutions. On the other hand, these criticisms suggest a desirable approximation of researchers from the commons with recent studies in political ecology, which in fact is already happening through researchers using the resilience framework (e.g., Beymer-Farris et al. 2012; Beymer-Farris 2013; Thorkildsen 2014; Ingalls and Stedman 2016; Hoque et al. 2017). This combination of perspectives would contribute to the restoration of the core values of social equity and justice of participatory biodiversity conservation through resistance to interventions from bureaucratic forces, donor-driven ideologies, and neoliberal influences (Dressler et al. 2010).

We have previously discussed that the notion (and associated ideology) of *development* was constructed during the age of modernity and that it has shaped our modes of social organization and relationships with nature (Chaps. 9 and 12). We have also emphasized that the demand for participation is not restricted to the popular claim or to a vanguard that dream about direct democracy at all levels, but is also linked to development policies and paradigms, which are based on industrialization, urbanization, and powerful technologies that aim to *master* nature (Chap. 9). Despite all its socio-environmental impacts, the current model of development based on economic growth seems to be strengthening, as well as the discourse associated with it.

Whenever we talk about economic growth, it is important to consider some reflexive (or provocative) questions, such as "who grew, how did they grow and at whose expense?". It is impossible to address these questions without mentioning the unbridled development of the global North, which relies on imports from the global South to help meet its rising demand for raw materials and consumer goods (Martinez-Alier and Milanez 2015). This pattern, in turn, has led to the existence of the *sacrifice zones*, an expression used mainly by environmental justice movements to designate places where a concentration of enterprises responsible for environmental damage is observed. Sacrifice zones are commonly found in areas with socially vulnerable populations, including traditional populations. Such areas might also have high levels of biodiversity, given the positive association between cultural diversity and biological diversity. The existence of sacrifice zones in both the global North and South is a consequence of the expanded reproduction of capital, which always requires new sources of accumulation that allow it to extend surplus value (Fig 13.6). The expansion of extractivism and the exploitation of labor through cheap and/or slave-like labor fulfill this strategic function (Leff 2006).

In this context, participatory biodiversity conservation cannot be based on corrections or technical adjustments to the current economic development model, but on the replacement of the subjacent logic with which it was built. In other words, the proposition of improvements to the business as usual model of development will not lead to different outcomes, neither will the numerous forms of accumulation by dispossession and environmental degradation diminish (Chap. 12, Büscher 2019). In this context, what would be necessary to ensure participatory biodiversity conservation? For Büscher (2019) what we need is revolutionary development beyond capitalism. A central aspect of this proposition is the right of societies to define and implement their own concepts and models of development. To this end, it is essential to construct an epistemic independence from the Eurocentric paradigms of modernity and its mental chains that perpetuate the cultural colonization that survived after territorial colonization (Mejía 2016). An emblematic example of an alternative and indigenous development model is the Buen Vivir—*Good Living* in Spanish—words used in Latin America to describe alternatives to development focused on the good life in a broad sense, but with the specific idea that well-being is only possible within a community and that nature forms part of this community (Gudynas 2011).

We have also previously presented the importance and moral merit of environmentalism of the poor for the conservation of biodiversity (Chap. 12). While not always presenting the best results in terms of number of species or density of rare species when compared to some fortress conservation experiments (Khadka and Nepal 2010;

Fig. 13.6 Protests demanding an agrarian reform and against the accumulation by dispossession model in Brazil. (Photos: The Pastoral Land Commission (CPT) archives)

Phalan 2018), this conservation model recognizes the relevance of the high levels of biodiversity in landscapes historically shaped by traditional populations (Chap. 3). However, both Buen Vivir and environmentalism of the poor (and even biocultural conservation) coexist unequally with the prevailing economic model, mainly due to the unlimited expansion of industrial extraction in the global South to meet the demands of the global North. For instance, despite the unquestionable advances in poverty reduction and significant social achievements of the leftist governments which incorporated the Buen Vivir in their constitutions (Bolivia and Ecuador), they have paradoxically adopted a neo-developmentalist model characterized by export-oriented extractivism which has resulted in innumerous socioecological impacts. This contradiction suggests there is little sense in believing in the success of alternative models to capitalism without the concomitant degrowth of the global North.

Degrowth is defined as a reduction of production and consumption at the local and global levels, and its main assumption is that human progress without economic growth is possible (Schneider et al. 2010). Despite degrowth being contrary to dominant paradigms such as neoclassical economics and Keynesian economics, it is not considered a new paradigm per se. Instead, it is constituted by a large array of concerns, goals, strategies, and actions where streams of critical ideas and political action converge (Demaria et al. 2013). Degrowth studies have multiple origins, constituting an expanding niche at the intersection of social and applied environmental sciences, comprising conceptual essays, empirical assessments, and, more recently, the study of its real-world implementations (Kallis et al. 2012; Weiss and Cattaneo 2017). Economic degrowth, far from being an option, is an inexorable trend, since, due to the depletion of natural resources and environmental destabilization of the conditions that prevailed in the Holocene, global economic growth rates have already begun to decline (Marques 2016). Accordingly, some authors stress that this transformation of models must be *sustainable*, in the sense of being environmentally and socially beneficial (Schneider et al. 2010; Martinez-Alier and Milanez 2015).

The proposed changes of sustainable degrowth in terms of economic and socio-ecological models are a source of inspiration for a new framework for biodiversity conservation: *convivial conservation* (Büscher and Fletcher 2019, Chap. 2). This framework is deeply related to participatory biodiversity conservation, as it considers local people as key decision-makers in conservation planning. However, it reverses the well-known Ostrom's polycentric governance, affirming that effective conservation governance should start by addressing actors in the upper class elite in order to target their actions and then work down toward the local people who manage the biodiversity so that pressures exerted on local conservation initiatives can be proactively addressed at their source. In the words of Büscher and Fletcher (2019): "merely focusing on local democracy without taking into account the power of 'outside' actors is naïve." While a biodiversity conservation model that promotes structural changes in the current power configuration might seem utopian, it appears to be the only alternative that could slow both biodiversity loss and socio-environmental injustices, which have been deepened with the rise of authoritarian right-wing governments.[3]

[3] I finished this chapter a few days after the military, police, and right-wing extremists in Bolivia forced President Evo Morales to resign and Jeanine Áñez, a Christian conservative with racist views, proclaimed herself the new president. Despite Morales's mistakes in his attempts to remain the president and the discontent of important unions and indigenous peoples with his government, there is no doubt that this was a civil-military coup, laden with brutal physical and symbolic violence. An emblematic example of such violence was the kidnapping and beating of Patricia Arce, an indigenous leader and the mayor of the municipality of Vinto, who had her hair cut forcibly, her body dyed red, and was subjected to a procession of miles without shoes and suffered multiple assaults. Besides this astonishing case, kidnappings, burning of houses of leaders of the party of Evo Morales, and several cases of rape of women by military and state police have been registered thus far. Most of these aggressions are accompanied by raised Bibles and the burning of the Wiphala flag— the flag of indigenous peoples—which denotes the religious, racist, and colonialist character of this coup. Moreover, it seems far from coincidental that this violence is directed toward indigenous peoples—those who have historically resisted and barred the privatization of the commons in this country.

In conclusion, throughout this book we have discussed indigenous peoples and local communities, environments, biocultural diversity, socio-environmental justice, human rights, popular participation, opposition to violence, and coexistence between different points of view, all cornerstones of participatory biodiversity conservation. The link between all these themes is undoubtedly democracy, despite its problems and challenges. The failures of the democratic process need urgent correction, but they should never be used as a pretext for the aggressions to democracy that are being made every day or even as an excuse to renounce it altogether.

Acknowledgments CB thanks the Conselho Nacional de Desenvolvimento Científico e Tecnológico (CNPq) for the Productivity Grant Fellowship (Process number 308628/2016-0).

References

Altman J (2003) People on country, healthy landscapes and sustainable indigenous economic futures: The Arnhem Land Case. The Drawing Board: An Australian Review of Public Affairs 4:65–82 https://openresearchrepository.anu.edu.au/handle/1885/76737

Apostolopoulou E, Drakou EG, Pediaditi K (2012) Participation in the management of Greek Natura 2000 sites: evidence from a cross-level analysis. J Environ Manag 113:308–318. https://doi.org/10.1016/j.jenvman.2012.09.006

Araral E (2014) Ostrom, Hardin and the commons: a critical appreciation and a revisionist view. Environ Sci Pol 36:11–23. https://doi.org/10.1016/j.envsci.2013.07.011

Arnstein SR (1969) A ladder of citizen participation. J Am Plan Assoc 85:24–34. https://doi.org/10.1080/01944363.2018.1559388

Azevedo W (2019) O zapatismo cresce e "rompe o cerco" In: Instituto Humanitas Unisinos. Available at: http://www.ihu.unisinos.br/591926. Accessed 20 Sept 2019

Bastakoti RC, Shivakoti GP (2017) Governing the commons through understanding of institutional diversity: an agenda for application of Ostrom's framework in managing natural resources in Asia. Redefining Divers Dyn Nat Resour Manag Asia 1:41–52. https://doi.org/10.1016/B978-0-12-805454-3.00003-7

Béné C, Newsham A, Davies M et al (2008) Review article: resilience, poverty and development. Annu Conf Hum Dev Capab Assoc New Delhi 623:1–30. https://doi.org/10.1002/jid

Berkes F (2004) Rethinking community-based conservation. Conserv Biol 18:621–630. https://doi.org/10.1111/j.1523-1739.2004.00077.x

Berkes F (2009) Community conserved areas: policy issues in historic and contemporary context. Conserv Lett 2:20–25. https://doi.org/10.1111/j.1755-263x.2008.00040.x

Berkes F (2017) Environmental governance for the Anthropocene? Social-ecological systems, resilience, and collaborative learning. Sustain 9. https://doi.org/10.3390/su9071232

Beymer-Farris BA (2013) Rethinking resilience through a political ecology lens: producing biodiversity in Tanzania's mangrove forests? In: Brannstrom C, Vadjunec JM (eds) Chapter 10 in Land change science, and political ecology and sustainability: synergies and divergences. Earthscan Publications, Oxford, UK

Beymer-Farris B, Bassett TJ, Bryceson I (2012) Promises and pitfalls of adaptive management in resilience thinking: the lens of political ecology, Chapter 15. In: Plieninger T, Bieling C (eds) Resilience and the cultural landscape: understanding and managing change in human-shaped environments. Cambridge University Press, Cambridge, UK

Bixler RP, Dell'Angelo J, Mfune O, Roba H (2015) The political ecology of participatory conservation: institutions and discourse. J Polit Ecol 22:164–182. https://doi.org/10.2458/v22i1.21083

Borrini-Feyerabend G, Pimbert M, Farvar MT, Kothari A, Renard Y (2004) Sharing power: learning by doing in co-management throughout the world. IIED and IUCN/CEESP/CMWG, Cenesta, London
Brondizio ES, Tourneau L, Francois-Michel (2016) Environmental governance for all. Science 352(6291):12. https://doi.org/10.1126/science.aaf5122
Brooks J, Waylen KA, Mulder MB (2013) Assessing community-based conservation projects: a systematic review and multilevel analysis of attitudinal, behavioral, ecological, and economic outcomes. Environ Evid 2:1–34. https://doi.org/10.1186/2047-2382-2-2
Brown K (2014) Global environmental change I: a social turn for resilience? Prog Hum Geogr 38:107–117. https://doi.org/10.1177/0309132513498837
Büscher B (2019) From 'global' to 'revolutionary' development. Dev Change 50:484–494. https://doi.org/10.1111/dech.12491
Büscher B, Robert F (2019) Towards convivial conservation. Conserv Soc 15:217–231. https://doi.org/10.4103/cs.cs
Carlsson L, Berkes F (2005) Co-management: concepts and methodological implications. J Environ Manage 75:65–76. https://doi.org/10.1016/j.jenvman.2004.11.008
Chilvers J, Kearnes M (2015) Remaking participation. Routledge, New York
CIMI (2018) Relatório Violência contra os Povos Indiǵ enas no Brasil – Dados de 2018 https://cimi.org.br/wp-content/uploads/2019/09/relatorio-violencia-contra-os-povos-indigenas-brasil-2018.pdf. Accessed 11 Dec 2019
Cornwall A (2008) Unpacking "participation" models, meanings and practices. Community Dev J 43:269–283. https://doi.org/10.1093/cdj/bsn010
Cote M, Nightingale AJ (2012) Resilience thinking meets social theory: situating social change in socio-ecological systems (SES) research. Prog Hum Geogr 36:475–489. https://doi.org/10.1177/0309132511425708
Cox M, Arnold G, Tomás SV (2010) A review of design principles for community-based natural resource management. Ecol Soc 15. https://doi.org/10.5751/ES-03704-150438
Cox M, Villamayor-Tomas S, Arnold G (2016) Design principles in commons science: a response to "Ostrom, Hardin and the commons" (Araral). Environ Sci Pol 61:238–242. https://doi.org/10.1016/j.envsci.2016.03.020
Dardot P, Laval C (2017) Comum – Ensaio sobre a revolução no século XXI. Boitempo
Davidson DJ (2010) The applicability of the concept of resilience to social systems: some sources of optimism and nagging doubts. Soc Nat Resour 23:1135–1149. https://doi.org/10.1080/08941921003652940
Demaria F, Schneider F, Sekulova F, Martinez-Alier J (2013) What is degrowth? From an activist slogan to a social movement. Environ Values 22:191–215. https://doi.org/10.3197/096327113X13581561725194
Demeritt D (2015) The promises of participation in science and political ecology. Routledge Handb Polit Ecol:224–234
Dressler W, Büscher B, Schoon M et al (2010) From hope to crisis and back again? A critical history of the global CBNRM narrative. Environ Conserv 37:5–15. https://doi.org/10.1017/S0376892910000044
Ellis EC (2019) To conserve nature in the Anthropocene, half earth is not nearly enough. One Earth 1:163–167. https://doi.org/10.1016/j.oneear.2019.10.009
Ellis EC, Mehrabi Z (2019) Half earth: promises, pitfalls, and prospects of dedicating half of earth's land to conservation. Curr Opin Environ Sustain 38:22–30. https://doi.org/10.1016/j.cosust.2019.04.008
EZLN - National Liberation Zapatista Army (2019) Comunicado del CCRI-CG del EZLN. Y rompimos el cerco. https://enlacezapatista.ezln.org.mx/2019/08/17/comunicado-del-ccri-cg-del-ezln-y-rompimos-el-cerco-subcomandante-insurgente-moises/. Accessed 13 Dec 2019
Gavin MC, McCarter J, Mead A et al (2015) Defining biocultural approaches to conservation. Trenin Ecol & Evol 30(3):140–145. https://doi.org/10.1016/j.tree.2014.12.005

Gavin MC et al (2018) Effective biodiversity conservation requires dynamic, pluralistic, partnership-based approaches. Sustainability 10(6):1846. https://doi.org/10.3390/su10061846

Godet L, Devictor V (2018) What conservation does. Tren in ecol & evol 33(10):720–730. https://doi.org/10.1016/j.tree.2018.07.004

Gruber JS (2010) Key principles of community-based natural resource management: a synthesis and interpretation of identified effective approaches for managing the commons. Environ Manag 45:52–66. https://doi.org/10.1007/s00267-008-9235-y

Gudynas E (2011) Buen Vivir: today's tomorrow. Development 54:441–447. https://doi.org/10.1057/dev.2011.86

Haggar J, Pons D, Saenz L, Vides M (2019) Contribution of agroforestry systems to sustaining biodiversity in fragmented forest landscapes. Agric Ecosyst Environ 283. https://doi.org/10.1016/j.agee.2019.06.006

Haklay M, Bowser A, Makuch Z, Vogel J (2018) Participatory citizen science. UCL Press:52–62. https://doi.org/https://www.jstor.org/stable/j.ctv550cf2.11

Hoque SF, Quinn CH, Sallu SM (2017) Resilience, political ecology, and well-being: an interdisciplinary approach to understanding social-ecological change in coastal Bangladesh. Ecol Soc 22. https://doi.org/10.5751/ES-09422-220245

Ingalls ML, Stedman RC (2016) The power problematic: exploring the uncertain terrains of political ecology and the resilience framework. Ecol Soc 21. https://doi.org/10.5751/ES-08124-210106

IPBES (2019) Summary for policymakers of the global assessment report on biodiversity and ecosystem services of the Intergovernmental Science-Policy Platform on Biodiversity and Ecosystem Services. Díaz S, Settele J, Brondizio ES et al (eds.). IPBES secretariat, Bonn, Germany. https://www.ipbes.net/global-assessment-report-biodiversity-ecosystem-services. Accessed 08 Dec 2019

Jose S (2012) Agroforestry for conserving and enhancing biodiversity. Agrofor Syst 85:1–8. https://doi.org/10.1007/s10457-012-9517-5

Kallis G, Kerschner C, Martinez-Alier J (2012) The economics of degrowth. Ecol Econ 84:172–180. https://doi.org/10.1016/j.ecolecon.2012.08.017

Khadka D, Nepal SK (2010) Local responses to participatory conservation in Annapurna Conservation Area, Nepal. Environ Manag 45:351–362. https://doi.org/10.1007/s00267-009-9405-6

Kohler F, Brondizio ES (2017) Considering the needs of indigenous and local populations in conservation programs. Conserv Biol 31:245–251. https://doi.org/10.1111/cobi.12843

Lane MB, Corbett T (2005) The tyranny of localism: indigenous participation in community-based environmental management. J Environ Policy Plan 7:141–159. https://doi.org/10.1080/15239080500338671

Leach M (2008) Re-framing resilience: a symposium report. STEPS Work Pap 13:1–18. https://doi.org/10.2307/977312

Leff E (2006) Racionalidade ambiental: a reapropriação social da natureza. Civilização Brasileira, Rio de Janeiro

López-Bao JV, Chapron G, Treves A (2017) The Achilles heel of participatory conservation. Biol Conserv 212:139–143. https://doi.org/10.1016/j.biocon.2017.06.007

Marques L (2016) Capitalismo e colapso ambiental, 2 ed ver e ampliada edn. Editora Unicamp, Campinas. ISBN 978-85-268-1337-3.

Martinez-Alier J, Milanez F (2015) Ecologismo dos pobres, Colonialismo e Metabolismo Social. InSURgência: revista de direitos e movimentos sociais 1(2):8–18

Matarrita-Cascante D, Sene-Harper A, Ruyle L (2019) A holistic framework for participatory conservation approaches. Int J Sustain Dev World Ecol 26:484–494. https://doi.org/10.1080/13504509.2019.1619105

Mehrabi Z, Ellis EC, Ramankutty N (2018) The challenge of feeding the world while conserving half the planet. Natu Sust 1(8):409–412. https://doi.org/10.1038/s41893-018-0119-8

Mejía MR (2016) Diálogo-confrontación de saberes y negociación cultural: ejes de las pedagogías de la educación popular: una construcción desde el sur. Educar em Revista 61:37-54. https://doi.org/10.1590/0104-4060.47205

Moote MA, McClaran MP, Chickering DK (1997) Theory in practice: applying participatory democracy theory to public land planning. Environ Manag 21:877–889. https://doi.org/10.1007/s002679900074

Ostrom E (1990) Governing the commons: the evolution of institutions for collective action. Cambridge University Press

Ostrom E (2009) A general framework for analyzing sustainability of social-ecological systems. Science 325(5939):419–422. https://doi.org/10.1126/science.1172133

Pateman C (2012) Participatory democracy revisited. Perspect Polit 10:7–19. https://doi.org/10.1017/S1537592711004877

Pearson LJ, (Lain) DM (2019) Framing up the "stretching" of co-management. Soc Nat Resour 32:363–381. https://doi.org/10.1080/08941920.2018.1544677

Pennington M (2012) Elinor Ostrom, common-pool resources and the classical liberal tradition. In: Ostrom E (ed) The future of the commons: beyond market failure and government regulation. Institute of Economics Affairs, London, pp 21–47

Perfecto I, Vandermeer J, Wright A (2009) Nature's matrix: linking agriculture, conservation and food sovereignty. Routledge

Perrings C, Jackson L, Bawa K et al (2006) Biodiversity in agricultural landscapes: saving natural capital without losing interest. Conserv Biol 20:263–264. https://doi.org/10.1111/j.1523-1739.2006.00390.x

Phalan BT (2018) What have we learned from the land sparing-sharing model?. Sustainability, 10(6), 1760. https://doi.org/10.3390/su10061760

Pretty JN (1995) Participatory learning for sustainable agriculture. World Dev 23(8):1247–1263. https://doi.org/10.1016/0305-750X(95)00046-F

Quinn CH, Huby M, Kiwasila H et al (2007) Design principles and common pool resource management: an institutional approach to evaluating community management in semi-arid Tanzania. J Environ Manag 84(1):100–113. https://doi.org/10.1016/j.jenvman.2006.05.008

Reyes-García V (2015) The values of traditional ecological knowl- edge. In: Martinez-Alier J, Muradian R (eds) Handbook of ecological economics. Edward Elgar, Cheltenham, UK, p 283

Rist P, Rassip W, Yunupingu D et al (2019) Indigenous protected areas in Sea Country: indigenous-driven collaborative marine protected areas in Australia. Aquat Conserv Mar Freshw Ecosyst 29:138–151. https://doi.org/10.1002/aqc.3052

Ross H, Grant C, Robinson CJ et al (2009) Co-management and indigenous protected areas in Australia: achievements and ways forward. Australas J Environ Manag 16:242–252. https://doi.org/10.1080/14486563.2009.9725240

Salomon A, Lertzman K, Brown K et al (2018) Democratizing conservation science and practice. Ecol Soc 23(1):44. https://doi.org/10.5751/ES-09980-230144

Schneider F, Kallis G, Martinez-Alier J (2010) Crisis or opportunity? Economic degrowth for social equity and ecological sustainability Introduction to this special issue. J Clean Prod 18:511–518. https://doi.org/10.1016/j.jclepro.2010.01.014

Singleton BE (2017) What's missing from Ostrom? Combining design principles with the theory of sociocultural viability. Env Polit 26:994–1014. https://doi.org/10.1080/09644016.2017.1364150

Sistla SA, Roddy AB, Williams NE et al (2016) Agroforestry practices promote biodiversity and natural resource diversity in Atlantic Nicaragua. PLoS One 11:1–20. https://doi.org/10.1371/journal.pone.0162529

Stephenson J, Berkes F, Turner NJ, Dick J (2014) Biocultural conservation of marine ecosystems: examples from New Zealand and Canada. Indian J Tradit Knowl 13:257–265

Tauli-Corpuz V, Alcorn J, Molnar A (2018) Cornered by protected areas: replacing 'fortress' conservation with rights-based approaches helps bring justice for indigenous peoples and local communities, reduces conflict, and enables cost-effective conservation and climate action. Rights and Resources Initiative, Washington, DC

Thorkildsen K (2014) Social-ecological changes in a Quilombola community in the Atlantic forest of southeastern Brazil. Hum Ecol 42:913–927. https://doi.org/10.1007/s10745-014-9691-3

Weiss M, Cattaneo C (2017) Degrowth – taking stock and reviewing an emerging academic paradigm. Ecol Econ 137:220–230. https://doi.org/10.1016/j.ecolecon.2017.01.014

White SC (1996) Depoliticising development: the uses and abuses of participation. Dev Pract 6:6. https://doi.org/10.1080/0961452961000157564

Wittayapak C, Dearden P (1999) Decision-making arrangements in community-based watershed management in northern Thailand. Soc Nat Resour 12:673–691. https://doi.org/10.1080/089419299279380

Yunupingu D, Muller S (2009) Cross-cultural challenges for indigenous sea country management in Australia. Australas J Environ Manag 16:158–167. https://doi.org/10.1080/14486563.2009.9725232

Zulu LC (2012) Neoliberalization, decentralization and community-based natural resources management in Malawi: the first sixteen years and looking ahead. Prog Dev Stud 12:193–212. https://doi.org/10.1177/146499341101200307

Zurba M, Ross H, Izurieta A et al (2012) Building co-management as a process: problem solving through partnerships in Aboriginal Country, Australia. Environ Manag 49:1130–1142. https://doi.org/10.1007/s00267-012-9845-2

Index

C. Baldauf (ed.), *Participatory Biodiversity Conservation*,
https://doi.org/10.1007/978-3-030-41686-7

The manufacturer's authorised representative in the EU is Springer Nature Customer Service Centre GmbH, Europaplatz 3, 69115 Heidelberg, Germany. If you have any concerns regarding our products, please contact ProductSafety@springernature.com

Printed and bound by CPI Group (UK) Ltd, Croydon, CR0 4YY
05/10/2026
02228124-0007